C000069430

The Business Occupier's Handbook

The Business Occupier's Handbook

A PRACTICAL GUIDE TO ACQUIRING, OCCUPYING AND DISPOSING OF BUSINESS PREMISES

Vicky Rubin

in association with

CLIFFORD CHANCE

Taylor & Francis

Taylor & Francis Group

LONDON AND NEW YORK

Taylor & Francis,
2 Park Square, Milton Park, Abingdon, Oxon, OX14 4RN
711 Third Avenue, New York, NY 10017

First edition 1997

© 1997 Victoria Rubin

Typeset in 10/12 Times by Saxon Graphics Ltd, Derby

First issued in paperback 2011

ISBN13: 978-0-419-21010-8 (hbk)
ISBN13: 978-0-415-51219-0 (pbk)

Apart from any fair dealing for the purposes of research or private study,
or criticism or review, as permitted under the UK Copyright Designs and
Patents Act, 1988, this publication may not be reproduced, stored, or
transmitted, in any form or by any means, without the prior permission in
writing of the publishers, or in the case of reprographic reproduction only in
accordance with the terms of the licences issued by the Copyright
Licensing Agency in the UK, or in accordance with the terms of licences
issued by the appropriate Reproduction Rights Organization outside the UK.
Enquiries concerning reproduction outside the terms stated here should be
sent to the publishers at the London address printed on this page.

The publisher makes no representation, express or implied, with regard
to the accuracy of the information contained in this book and cannot
accept any legal responsibility or liability for any errors or omissions
that may be made.

A catalogue record for this book is available from the British Library

Contents

Foreword

This book is designed to fulfil the need of every business occupier for a comprehensive but approachable source of information about the numerous issues which need to be understood in order to play an effective role in acquiring, managing, improving and disposing of business property.

It brings together expertise from a wide range of disciplines – surveyors, accountants, valuers, construction cost and project managers and lawyers – set down in a practical and non-legalistic style by a chartered surveyor with direct experience of property management. Its aim is to deal with each topic in sufficient detail for the purposes of the business reader, concentrating at each step on the commercial as well as the technical considerations.

Ready access to this level of information will help the business occupier in dealings with professional advisers, agents, property managers, contractors and others, and contribute posititvely to the decision-making process. We believe that the advisers, managers, contractors and agents will also find much of use in the handbook.

We are very pleased to have been able to support and contribute to this project and we hope that the handbook will come to be seen as a handy and indispensable ready reference by everyone who owns and uses it.

Clifford Chance
August 1996

Acknowledgements

I would like to thank all those at Clifford Chance who gave their support. From the beginning the firm embraced the project with enthusiasm and willingly devoted to it the resources needed. In particular, those who have devoted considerable time are: Julian Boswall, Teddy Bourne, Tony Briam, Richard Coleman, Steven Dale, Alexandra Davies, Alan Elias, Trevor Garrood, Graham Henderson, Clive Jones, Vincent Maguire, Richard Margree, Shelagh O'Brien, Tim O'Hara, John Pickston, Michael Redman and Jeremy Sandelson.

Special thanks are due to the following who contributed text and provided vital technical support:

- Andy Simmonds of Deloitte and Touche (chartered accountants);
- Nigel Laing of Nigel Laing and Co (specialist valuation surveyors);
- Beverley Stewart of Allen Stewart Partnership (construction cost and project managers).

I would also like to give special thanks to Simon Brooke, Jonathan Edwards and Michael Evans of Jonathan Edwards Limited, who provided crucial inspiration and education while the book was in gestation and assistance in preparing the text.

I am indebted to several others who gave invaluable help: Sandy Apgar of Apgar and Company, Simon Barnes of Plowman Craven, Michael Bloom of Business Automation, Edward Clack of Edward Clack Associates, Malcolm Hull of Drivers Jonas, Stuart Johnson of Watts and Partners, Adrian Leaman, Rebekah Lowe of the RICS, Trevor Rushton of Watts and Partners, Katherine Ryan of Jonathan Edwards Limited, Nick Schroeder of Rank Xerox, Steven Sheppard of Segal Quince Wicksteed, Martin Trundle of Jonathan Edwards Limited, Howard Tucker of Allen Stewart Partnership, Chander Vasdev of Rank Xerox and Mary Williams of KPMG.

Finally, I would like to thank John Penny for his inspired illustrations.

Introduction

Property is a key element in the identity of most firms. As the environment in which the business operates, it can condition the way the business functions. Property costs also have a direct impact on profitability: the cost of occupational property is second only in size to staff costs. Effective property management can substantially affect occupancy cost and in turn profitability.

And yet companies do not always appreciate the influence of property on corporate success. They often fail to apply the resources – time and money – which reflect the significance of property for the business. Many businesses are paying more than they need to for their property, have too much space or too little, are located in the wrong place, or in the wrong type of building.

A portfolio left to run itself will not operate efficiently. Strategic and detailed management are essential to effectively managing cost and achieving a property portfolio which facilitates – rather than obstructs – the core business. By applying the principles and using the information explained in this book you will be in a position to negotiate better transactions, deal more effectively with landlords, satisfy occupational needs, control cost, and get the most from your professional advisers. The result will be a property portfolio which meets the needs of the users – and satisfies the finance director!

When I was managing investment properties for landlords, I regularly encountered tenants who did not appear to understand the legal framework governing the landlord and tenant relationship, and who, as a result, failed to protect themselves properly. Subsequently, the time I spent advising occupiers confirmed my view that a book was needed to provide the business occupier with the technical framework and strategic advice needed to manage corporate property effectively.

The book aims to bring together a wide range of issues in order to provide the occupier with the basic framework relevant at all stages of occupation. It cannot provide an exhaustive discussion on all these issues; if you wish to understand a particular area in detail you will need to refer to a specialist text. In particular, this is not a guide to facilities management, a detailed discussion on day-to-day operational issues; the book touches on operational issues, but focuses primarily on the strategic business, legal and financial framework. Books which provide more detail on operational issues and other specialist areas are recommended in the Further Reading sections at the end of each chapter.

It has been my aim to meet the needs of both small and large businesses, but to some degree these are contradictory. Many small occupiers, with only one or two occupational properties, may not wish to concern themselves with all the technical detail – at least until they encounter a particular problem that needs solving. Larger occupiers who deal with property issues on a more regular basis can make greater use of the technical information. A word of caution, however. All too often it is the small businesses that are worst protected in the jungle of the commercial property world – despite being the least able to afford it. The managers of small businesses will benefit from taking the time to read the substance of the book; it will provide you with the background necessary to minimize costs and protect yourselves effectively.

The book is not intended to be a substitute for professional advice. The technical information should only be read as general advice, providing a starting point for understanding a particular issue; you should always take professional advice to establish the precise position concerning the issue at hand. The book will however assist you in determining when advice is necessary, enable you to get more from your professional advisers and help you deal more effectively with the issues on which advisers are not instructed.

The book covers the law and practice in England and Wales and, while much of the practical framework will be similar in Scotland and Northern Ireland, the technical details, and particularly the legal position, will be different. The law is as it stands at the time of going to print in August 1996. The advice contained here reflects the issues that are relevant in the current market although many of the strategies will be appropriate if market conditions change.

Where abbreviations are used in the text they are normally defined, except in some cases, such as some professional bodies including the RICS and ISVA, where they are used frequently and are referred to in the addresses section at Appendix C.

Finally, throughout the text I have used the male form and avoided 'he or she' references. In all cases the male should be read as including the female. I believe this makes for a more readable text, and hope the reader will forgive this departure from political correctness.

Vicky Rubin

<table>
<tr><td>

Strategic and background issues

</td><td>

1

</td></tr>
</table>

KEY POINTS

- Impact of strategic property management on profitability
- Key issues in effective property management
- Analysing options for effective property decisions
- Getting the most from advisers and agents
- Instructing valuers and understanding valuations
- Accounting and tax issues for property decisions

This chapter covers the general issues affecting property at all stages of the process, starting with property strategy itself and then covering decision making, using advisers, valuation, accounting and tax. These are all important areas that the occupier is likely to encounter at some stage in the process, and possibly on a regular basis. The chapter provides the broad technical framework to enable the occupier to take an intelligent informed approach to these issues as they arise, and to know when an issue should be addressed in detail. The occupational process, starting with acquisition, and moving on through occupation to disposal, is discussed in detail in Chapters 2 to 5.

1.1 PROPERTY STRATEGY

Despite the importance of operational property for corporate profitability, it has not generally been given the attention it deserves by senior management. Property has traditionally been neglected in the boardroom and while there has been some improvement in recent years, many senior managers still fail to recognize the impact it can have on corporate profitability. As recently as 1992 Debenham Tewson Research reported the results of a survey of 100 major UK non-property companies, saying:

Only on rare occasions does property receive explicit treatment in corporate business plans. More often than not property is viewed as incidental

Property usually represents a business's largest asset and second largest cost after payroll. Cost savings translate almost directly into profit and the impact of strategic property management on profitability and share value can be significant. The neglect of property by senior management cannot be explained by lack of importance; instead it probably relates to the perception that property costs are outside the control of the business, driven by the market. In the past, the lack of interest in property has largely been a consequence of complacency resulting from the inflationary environment in which rising property costs have been obscured, and of the dominant perception that rising property costs are an inevitable corollary of growth. In recent deflationary times, with all costs coming under closer scrutiny, property costs have been the subject of particular concern. The perception that they are fixed and uncontrollable has however persisted in many quarters and there is often still a failure to recognize the scope for occupancy cost management.

In fact there is significant scope to control occupancy cost. A few notable examples have demonstrated the potential for reducing cost. Rank Xerox, for example, recently cut their occupancy cost by 20% in a two-year period, by a combination of disposals and efficiency initiatives. Smaller companies may not have quite the same scope for economies, but significant savings can still be made and in a small company these can sometimes have a proportionately larger impact on profitability.

Cost is however only one side of the corporate property equation. Property is a factor of production and the role of corporate property is as a support service designed to satisfy the needs of the core business units. Corporate property strategy must satisfy both operational and financial demands. The dual objectives of cost control and performance are often in conflict and it is the role of the business manager to achieve the optimal balance between the two.

The issues involved in strategic corporate property management are complex and a detailed discussion is not within the scope of this book. Please refer to one of the specialist texts on the subject (Further Reading), such as Nourse (1989) or Silverman (1987). The next sections focus on some of the central issues, which business managers should consider as a starting point for developing property strategy.

1.1.1 The link with business strategy

A coordinated property strategy is essential to effective management of the portfolio in accordance with business needs. Even small companies with only one or two properties need to consider their business requirements and translate these into a defined strategy. At the most simple level this may be establishing a need to relocate, or to dispose of a surplus building. For larger organizations the options available are more numerous and the analysis and debate involved in agreeing the strategy are more complex.

Figure 1.1 shows the process involved in developing property strategy.

The key here is the link between corporate and property objectives. The process of translating corporate objectives into a property strategy is crucial to ensuring that property decisions are driven by business strategy. The alternative is a haphazard approach to property decision making driven by individual business managers without explicit objectives or a coordinated strategy. The property strategy developed forms a strong foundation for specific property decisions, such as whether to buy or lease, choosing between different locations and properties and deciding whether to dispose of premises. These types of issue are discussed in detail in the following chapters.

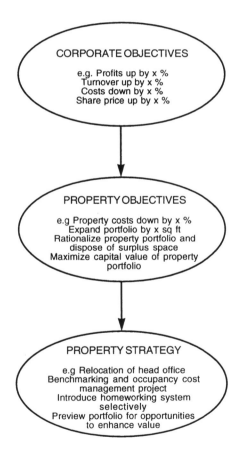

Figure 1.1 Developing property strategy.

By articulating corporate objectives and translating these into property objectives and strategy you achieve a property strategy consistent with corporate goals. The process of explicitly identifying objectives and analysing and debating options enables management to contribute to the decision-making process and arrive at a strategy agreed and supported by all. By establishing explicit goals and methods, the different parts of the organization can work cohesively with a coordinated focused approach.

The role of property in corporate profitability needs to be recognized by senior management and property given a central place in decision making. Property specialists need to highlight the link between corporate goals and property decisions and become better integrated into the corporate planning process.

1.1.2 Strategic property management

Active management can transform the property portfolio. A sales organization halves its property holdings by implementing teleworking and desk-sharing systems; a professional firm finds it can reduce cost and improve the working environment by a move towards a branch office portfolio instead of concentration of staff in a central London head office; a retailer improves sales substantially following a programme of disposals and acquisitions, achieving a portfolio of outlets distributed in accordance with current consumer demand patterns. These are examples of the improvements that can be made by taking an active approach to property management.

There are three objectives that normally motivate strategic property management:

- meeting operational needs – providing the space required by the core business in terms of the size, type, condition and specification of space, its location and the leasing or purchase terms;
- meeting occupancy cost objectives;
- enhancing and releasing asset value.

Capital value can have an important effect on the balance sheet. Value enhancement can often be achieved using some of the following strategies.

- sale and leaseback transactions;
- redevelopment (with or without a development partner);
- acquisitions and disposals.

However, for most occupiers asset value is secondary to the primary rationale of serving the core business units effectively at minimum occupancy cost. Traditionally, managers responsible for dealing with property decisions have tended to apply investment appraisal techniques in analysing property decisions, because the investment role of property has often been the dominant concern. For most occupiers now the crucial issues are occupational need and affordability, not investment value and capital appreciation.

The aim is to manage property costs to fit with the business plan. This requires considered explicit decisions on the type, quality, location and quantity of space occupied, in contrast to maintaining a portfolio whose size and structure is justified only by historic decisions. Occupancy cost management is considered in detail in the next section.

(a) Occupancy cost management

In general, the objectives of occupancy cost management are twofold:

- Cost efficiency – to achieve the lowest cost for a given level of service. This involves measuring occupancy cost, comparing it with other occupiers and analysing processes and practices to improve performance.
- Cost/performance trade off – to achieve the most appropriate balance between cost and level of service. Performance relates to the quality of the space and facilities, the facilities management service and the extent to which the facilities meet the needs of the business users. A key issue is flexibility which enables property managers to accommodate changing business needs.
  In many companies the standard of performance is unquestioned; the relationship between facilities performance and profitability is regularly overlooked and the cost/performance balance is not considered or debated explicitly. By analysing cost choices with an open mind corporate efficiency and profitability can be improved.

Figure 1.2 illustrates the relationship between the dual objectives of efficiency and the cost/performance balance. In this example, Company A needs to close the efficiency gap to move from point X onto Company B's trade-off line. Once Company A achieves this trade off they may not choose Company B's point on the line, Y, as they may decide that a different balance between cost and performance, for example Z, is appropriate to their business needs.

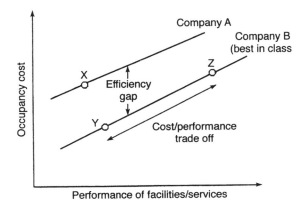

Figure 1.2 Occupancy cost management – objectives.

The corporate property manager needs to identify the organization's preferred cost/performance balance in addition to achieving the lowest level of cost for a given level of performance. In practice, management of occupancy cost to achieve efficiency happens during occupation and many of the issues relevant to this process are discussed in Chapter 4; choices relating to the quality and nature of the space are in practice made as part of the acquisition and disposal processes, which are discussed in Chapters 2 and 5 respectively. The decisions are integrated into the practical process of finding and acquiring space and subsequently deciding to dispose of it. The practical aspects of these processes are discussed in detail in the following chapters, but in order to manage those processes effectively the underlying objectives and priorities of the organization need to have been established.

It is important to understand clearly the factors that drive occupancy cost, both generally and for your particular organization. A useful framework has been developed by a US management consultant, Mahlon Apgar. He highlights 'The Three Ls', location, layout and leasing, as the factors which determine occupancy cost levels.

- Location. The supply and demand relationship in the property market is location specific and all occupancy costs are affected by location to a greater or lesser extent. This is most evident in the case of rent and property taxes (which in the UK are rent-related) but also applies to other costs to the extent that the economy generally exhibits locational cost differences. In the UK non-rental cost differentials are clearly demonstrable, for example between town centre and out of town locations, north versus south and so on. Arguably locational cost differentials may diminish as technological change increases the dispersion of business activity.

  For the occupier the challenge is to ensure that every activity within the organization is located in the most cost-efficient location taking into account the needs of the business users. In practice it is not possible to achieve a perfect distribution of all activities in the most cost-efficient location and this can only be approximated. It is important however that the occupier recognizes the impact of location on cost and analyses corporate needs and objectives to choose the most appropriate location for all areas of the business. In some cases economies achieved by consolidation may outweigh savings achieved by a dispersion of activities according to locational needs and the consolidation/diversification question needs to be considered as part of the locational decision.

- Layout. Layout is the amount and type of space used by the organization as well as the manner in which the space is used and located internally. The most significant issue is usually the quantity of space. This is often seen as outside the control of the property manager and driven by the business units or by historic levels of space usage. It is important to recognize that the size of the property portfolio can be chosen based on analysis: space can be acquired and disposed of in the medium to long term.

Use of space should be seen as an investment decision. Breakeven analysis and net present value analysis (Section 1.2) can assist in evaluating expansion and contraction. In all cases the space usage decision should take into account both revenue and cost: the use must be justified not just in terms of relative costs but in terms also of the revenue that will be generated from the space. Managing growth and controlling demand is an important role for the strategic property manager. Internal rent charging systems (discussed below) can assist in this process by linking the demand for space to profitability.

The quantity of space used is affected by working practices and systems. Space usage can often be dramatically reduced by space efficiency initiatives such as teleworking and desk sharing ('hot desking') for offices, and just-in-time delivery systems and centralized distribution centres for retail premises. In the long run changing working practices are likely to transform the size and nature of demand for space.

Layout refers not only to the total quantity of space used but to the nature and pattern of usage. The arrangement of uses needs to reflect the cost differentials between different areas within buildings as well as the relationship between uses. Additionally, the quality of the space, its specification and condition, should be designed to maximize productivity. For office space, the specific relationship between the amount and quality of the space allocated to a member of staff and that staff member's work output can be hard to measure. In retail space the impact of the quantity and quality of space on revenue can be more easily established from sales data.

- Leasing. The transaction achieved at the time of an acquisition or at a subsequent alteration of the terms, such as a rent review is the third cost driver. In theory this should be the least important issue because in a competitive economy changes in leasing costs should correlate with changes in the occupier's revenue. In practice, the economy does not adjust perfectly and fluctuations in the market can severely affect profitability.

While as an occupier you cannot control the supply/demand relationship in the property market, you can manage the property portfolio and control the acquisition and disposal processes to achieve the most efficient transactions on the most favourable terms available. A full analysis of options and effective negotiation to achieve the most appropriate terms to meet business needs are central to the process. These issues are discussed in detail in Chapters 2 and 3.

The Three Ls is an effective model for understanding occupancy cost; it is a useful starting point for making choices about the level of occupancy cost and configuration of space. The framework can be used as a basis for analysing the portfolio, identifying areas for change and formulating a coordinated strategy for property in line with corporate objectives.

(b) Information

A vital ingredient of effective property management is a comprehensive and accessible database of all relevant information about the portfolio. It is only by having to hand full information about the current position that problems and opportunities can be identified. By understanding the portfolio thoroughly, you will be able to assess how far it meets business needs in terms of cost, operations and value, and identify areas for change.

The system will depend on the use to which you intend to put the information. An operational system used for day-to-day management, such as payment of rent and other charges, diarizing rent reviews, lease expiries and so on, may not be the best means of organizing information for strategic management. The basic information stored on the database should include:

- premises – location, floor area, specification, condition, facilities, use, and other occupiers;
- transaction – tenure, length of term and expiry/break dates, rent reviews and other terms;
- occupancy cost – rent, rates, service charge, operating expenses and depreciation;
- rental and capital value – book value, current estimated rental and capital value, last valuation date, planning permissions, redevelopment and other issues which affect value.

However, the value of the database will be enhanced if business information is also recorded to assist in performance measurement. Details for each property of staff numbers, turnover, profit levels, floor area per employee and occupancy cost per employee, will all assist in effective strategic property management.

Before property costs can be recorded, analysed and managed effectively they need to be accurately measured. This depends on an appropriate and consistent definition of occupancy cost and the various categories within it. The first step is to decide on which costs should be included within the definition. The American National Association of Accountants (ANAA) produced a statement *'The Accounting Classification of Real Estate Occupancy Costs'* (January 1991). In the absence of an agreed UK definition, this provides a useful definition of occupancy cost as the total cost incurred by the company to provide space for the operations of its business units. It suggests four tests for identifying whether a cost should be included within occupancy cost:

- Does the cost directly reflect a cost of providing, maintaining, or using the real estate?
- Does the cost reflect a standard or typical amount or type of cost that would be charged, passed-through, or used as a basis for determining the rent for space leased by the company from an independent third party?
- Is the type of cost one that would be incurred by a typical user, or does it reflect special requirements of the business unit or the company? Any

unusual cost which reflects the special requirements of the individual business should be reflected as expenses of conducting business rather than as occupancy costs.

● Is the type of cost one that the company would (or could) charge if it were to lease space to a typical unrelated firm?

The categories of occupancy costs are not always defined on the same basis and it is important when analysing and comparing costs to ensure that they are grouped together in a consistent way. Figure 1.3 shows a commonly used basis for categorization of occupancy costs. The first four categories are usually cashflow figures, except in the case of notional rents which are normally a cost of capital used to represent the cost of owned properties. Depreciation is an accounting charge.

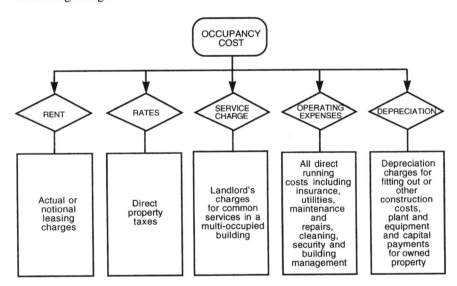

Figure 1.3 Occupancy cost categories.

Rent, rates and service charge are clearly distinguishable as they are charged by the landlord. Operating expenses and depreciation are not always as simple to quantify because it can be difficult to determine which costs should and should not be included. A fit-out contract for office space, for example, will commonly include a range of items including base building works, decorative finishes, fixtures, fittings, furniture and IT equipment. Some of these costs, particularly fittings and furniture and some IT hardware, do not pass the ANAA tests, but it is often hard to distinguish accurately those that do. Should demountable partitioning be included? Computer hardware? Computer cabling? Hard furnishings? Soft furnishings? This is a grey area and a consistent basis has not been established. It is important that your analysis of your own and other companies' occupancy cost takes into account any differences in the measurement basis.

Definitions of occupancy cost are only the first step to establishing a system for measuring it. Internal systems need to be developed including coordinated systems for recording invoices and payments. Measuring and reviewing occupancy cost should be undertaken regularly as part of the management accounting and budgeting process. It is helpful to track costs over a period and prepare an occupancy cost history. This can identify a pattern over time as a first step in the analysis before moving on to internal and external comparisons as discussed below under Property Benchmarking.

(c) Management systems

Turning to the specifics of property management, the first issue that should be considered is the system of management of the property portfolio and in particular the method and level of centralization of decision making. Corporate property is commonly managed centrally by a property department or property holding subsidiary but the degree of autonomy exercised by the operational departments varies. In general the aim is to give the operational departments or local branches enough control to ensure that specialized needs are met while at the same time maintaining a coordinated approach guided by an overall strategy. The specific balance achieved in a particular case depends on the nature of the business and the size of the portfolio.

The other vital element of a successful corporate property management system is a mechanism for linking the space usage and cost decision at the operational level. In many companies space needs can only be identified by the operational units. Operational units, however, tend to have unlimited demands and can fail to evaluate their space requirements against cost objectives. There is a need to avoid a situation in which the operational units produce 'wish lists' which are then policed by the central property department.

Some organizations have developed systems for charging occupancy costs to operational units based on space usage, in some cases using sophisticated computer aided design (CAD) technology to monitor usage and assign cost levels. Under this type of system the profitability of the operational units is affected by their space usage and space decisions are intrinsically linked to cost concerns. This type of internal rent system can have significant advantages in achieving an efficient use of property as well as assisting in accurate product pricing. The main disadvantages are the time and cost involved in establishing and running the system and the potential for engendering an overly competitive environment within the organization.

(d) Property benchmarking

Benchmarking is a business management tool pioneered by Rank Xerox in the early 1980s. Their chief executive David Kearns defined benchmarking as 'the continuous process of measuring products, services, and practices against the toughest competitors or those companies recognized as industry leaders.' The concept is simple: to learn from the experience of others. Rank Xerox used the process to pare down costs and enhance profitability. In the realm of occupational property it can be used to evaluate and improve property management techniques.

A full benchmarking project should address both quantitative and qualitative issues. The quantitative issues, cost and other financial performance measures, tend to be the focus of much benchmarking work, but non-financial performance measures are equally important. In the case of the property function the qualitative issues relate to the quality of the services provided to business users.

There are four elements to the process:

- Internal benchmarking – comparing costs, processes and practices within the organization. The aim is to look at cost histories and for departments to learn from each other to ensure that all areas of the business reach the level of the most efficient ones. Processes and methods can often be translated from one part of the business to improve performance in another. Internal benchmarking is often the most constructive as it frequently facilitates more meaningful comparisons than are possible with the following three, external, methods.
- Competitive benchmarking – examining how competitors do things and learning from them.
- Industry benchmarking – examining other companies within the industry.
- Best in class benchmarking – identifying high achieving companies in a particular product, function or performance area and learning from them. A company may be picked because of its attainment of a particular performance standard, such as their occupancy cost to turnover ratio or another relevant measure; or because of efficiency in performing a particular function, such as credit control, personnel or facilities management. In some cases, the company can be chosen on the basis of a specific issue, such as low energy or maintenance costs, where this is the focus of investigation.

The quantitative issues are the most easily addressed but to focus on these in isolation can be dangerous. Keeping in mind the dual objective of efficiency and the best cost/performance balance, it is important to evaluate the quantitative issues together with qualitative performance criteria. To take an example, focusing on rental costs in the absence of an analysis of space quality issues can result in a distorted result. Meaningful cost comparisons can only be made with standard of the product in mind. The aim is to achieve value for money taking into account the standard of facilities provided. Having addressed the efficiency issue, the company can turn to the cost/facilities trade-off decision and consider what level of performance is required.

The first step in benchmarking is to define costs and performance and undertake an internal benchmarking study to analyse differential performance within the organization. It is then appropriate to move on to external benchmarking. Figure 1.4 shows the steps involved in both internal and external benchmarking projects.

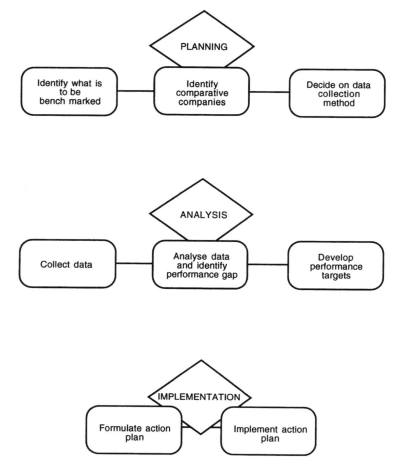

Figure 1.4 Benchmarking process.

The project can relate to the business as a whole, in which case property will be a part of the study, or may focus specifically on the property management function or a particular element of it. In each case both macro and micro performance measures can normally be used. In the case of a business-wide study there will usually be an emphasis on the macro level analysis, while a function specific survey may focus on detailed micro level measures, such as specific cost headings.

Table 1.1 lists examples of quantitative performance measures commonly addressed in a property benchmarking project. The quantitative measures which address output, such as revenue and profit per square foot or per square metre, go some way to evaluating performance as well as cost. A comprehensive benchmarking project however will also analyse performance using qualitative performance measures.

Table 1.1 Benchmarking quantitative performance measures

Macro level performance measures	Ratio of occupancy cost to profit
	Ratio of occupancy cost to turnover
	Ratio of occupancy cost to total expenses
	Revenue per square foot (per square metre)
	Occupancy cost per square foot (per square metre)
	Profit per square foot (per square metre)
	Floor area per employee
	Occupancy cost per employee
Micro level performance measures	Rent
(cost per square foot or square metre)	Rates
	Service charge
	Leasehold depreciation
	Property management
	Insurance
	Cleaning
	Energy
	Security
	Building maintenance
	Plant maintenance
	Catering
	Office services

Having benchmarked cost and quality the study next needs to turn to processes. A property benchmarking project should not focus exclusively on performance measures but should also review and compare property management processes. These include strategic decision making, space planning, estate and facilities management, outsourcing, landlord liaison, cost audit and invoice approval systems.

Out of the comparison the gap between the company's performance and best practice can be identified and performance targets and process changes developed. For this to be done effectively it is important that goals and methods are debated. Some performance targets and processes identified from the benchmarking project will not be appropriate or may be impractical to adopt. The company may choose a different trade off between cost and quality than the benchmarked company and the performance targets adopted must reflect this decision. For example the benchmark partner may decide that top specification offices in a prime location are necessary, whereas you may decide on a lower quality of space. If the most effective balance to meet business needs is to be achieved, all relevant business managers should be involved in the decision.

Benchmarking is an ongoing process. Its success depends on the investment of time and money and it will only be successful if the following conditions are satisfied:

• an active commitment to benchmarking from management;

- the full involvement of all business managers in the analysis, decision making and implementation processes;
- a culture receptive to innovation and an openness to new ideas within the organization generally;
- a willingness to change and adapt based on benchmarking findings;
- a willingness to share information with benchmark partners.

A property benchmarking study may be undertaken as a stand alone project or as part of a general benchmarking study. In many cases it is advisable to manage the project with the help of a specialist benchmarking consultant. This will depend on the in-house expertise and resources available. In the case of a macro level study a management consultant will normally be appropriate. Where property is the only or a significant focus of the analysis it is usually advisable to instruct a specialist property benchmarking adviser, sometimes as part of a wider team.

A list of consultants providing benchmarking services is included in Appendix C. This provides a starting point for occupiers considering benchmarking but you should undertake your own research to find the most appropriate service to meet your needs. An occupier may choose a full-blown project managed by a consultant involving one or more benchmarking partners, or it may use a streamlined database route, or a combination of the two. The database approach is usually more economical and provides a range of benchmarking data cost effectively, but it does not facilitate a comprehensive analysis of processes. Usually the most cost-effective approach is to subscribe to a benchmarking database with workshop and advisory services, to enable a combination of financial evaluation and analysis of processes.

This section has addressed cost at a strategic level and has looked at the underlying issues that drive occupancy cost. These issues are explored in more detail in the following chapters, and in particular the decision-making process for acquisitions and disposals is covered in Chapters 2 and 5 respectively. Management of occupancy cost on a detailed level is addressed in Chapter 4 where each of the detailed components – rent, rates, service charge, operating expenses and depreciation – is considered in detail.

Many of the issues discussed here are complex to implement in practice. The occupier should take the advice of one or more experienced professionals in the field. These may be a management consultant, surveyor, property adviser or management accountant, depending on the type of input required. In view of the potential impact on profitability, the time and cost involved in addressing these strategic issues comprehensively are usually justified, although they will need to be tailored to the size of the portfolio and the level of potential savings.

1.2 DECISION MAKING

Property decisions usually have a long-term impact on the business and its profitability. In any decision-making situation it is essential that options are analysed rigorously to ensure the right decision – the one which best meets business needs – is chosen. Whether it is a strategic decision, such as consolidation of the property portfolio, or a specific transaction such as an acquisition or disposal, all aspects of the decision need to be analysed to determine how far it meets the objectives of the company. A full evaluation of all elements of the transaction enables decisions to be made effectively on the basis of information and analysis rather than partial judgement and gut reactions.

This is not the forum for a detailed discussion on investment appraisal and financial analysis techniques. Please refer to a specialist text for a full treatment of the subject – Lumby (1991), Mott (1993) and Levy and Sarnat (1990) all provide good discussions. This section can only touch on the subject but aims to provide a structure for appraising property transactions.

Rigorous analysis assists in decision making by helping the facts to speak for themselves. The occupier may need to evaluate one particular transaction, or to compare a number of options. In each case financial analysis provides the tools to facilitate the decision. The evaluation should normally include three elements:

- Economic analysis to evaluate the cashflow impact of the transaction. This will normally comprise discounted cashflow analysis to arrive at a net present value (NPV) figure that can be compared easily with other options. In some cases a separate analysis of risk is also necessary;
- Accounting analysis to evaluate the impact of the option on the company's financial statements;
- Qualitative and quantitative analysis of the wider business impact of the decision. This is an evaluation of the impact of the decision on the business as a whole. The aim of this part of the analysis is to ensure that the decision is evaluated in its widest context taking into account all its implications for the business.

From these three elements of the review the manager will have the ammunition to make an effective decision. Each is considered in turn below.

1.2.1 Economic analysis

Appraisal methods are plentiful and the method best adopted will depend on the objectives to be achieved and the other circumstances. In some situations it will be appropriate to consider a number of measures in addition to net present value (NPV), such as the payback period or the internal rate of return (IRR). In most cases however the NPV analysis will be the most appropriate starting point for the appraisal and other methods will be complementary. This section focuses on discounted cashflow analysis used to determine the net present value as well as touching on risk analysis which can be relevant in some cases. For a more detailed discussion on these and other appraisal techniques please refer to one of the specialist texts listed in the Further Reading section.

(a) Net present value analysis

The NPV calculation quantifies all the expected cashflows, income and expenditure, over an appropriate period. The analysis enables options to be compared by converting their cashflows into present value equivalents. Spreadsheet computer packages facilitate the calculation relatively quickly and easily.

The period used for the analysis depends on the nature of the transaction; where, for example the transaction being evaluated is the acquisition of a lease, the period will normally be the lease term. The cashflows are discounted to their present-day equivalent, to reflect the time value of money.

The discounted cashflow will normally be focused around the heads of expenditure included within occupancy cost – rent, rates, service charge, operating expenses and depreciation. Other costs and receipts should also be considered where relevant. These may include tax costs or savings, transaction costs such as estate agents', valuation and legal fees and any cash receipts such as capital payments or income from subletting.

Example 1.1 A company has acquired new office premises but still holds an old office building on a lease expiring in three years at a passing rent above the current market level. The company has approached the landlord and has arranged to meet with him to discuss a surrender of the lease. The company wishes to determine the maximum price they should pay.

Table 1.2 shows the discounted cashflow analysis used to determine the net present value of outgoings under the lease if it is left to run until expiry.

Table 1.2 Discounted cashflow analysis to evaluate surrender premium

	Net present value	1996	1997	1998	Total
Rent	£1 280 228	£464 000	£464 000	£464 000	£1 392 000
Rates	£200 634	£70 056	£72 860	£75 722	£218 638
Maintenance	£413 867	£150 000	£150 000	£150 000	£450 000
Dilapidations	£84 168			£100 000	£100 000
Total	£1 978 897	£684 056	£686 860	£789 722	£2 160 638

Discount rate: 9%.

The result in this case shows that the company should pay a maximum of £1 978 897 in a surrender payment to the landlord, based on the assumptions made, including the assumption that a subletting will not be possible because of the short period left on the lease. If the company is able to negotiate a lower premium the surrender will be advantageous.

The discount rate used in the analysis must reflect the cost, or opportunity cost, of funds invested. The figure normally used is the company's weighted average cost of capital, but in some cases an opportunity cost rate, such as the return that would be achieved from funds invested in the core business, may need to be adopted.

The analysis is relatively simple to prepare using a computer spreadsheet package which calculates the NPV of a series of cashflows. The difficulty arises in ensuring that all potential costs are considered and accurate assumptions made where figures have to be estimated. You will need to take into account all factors that may affect the cashflow, such as rent reviews, rating revaluations, other tax issues, repair and maintenance obligations and service charge expenditure.

The quality of the results achieved are dependent on the quality of the assumptions. Taking good professional advice is usually essential if the NPV figure is to be reliable. It is often advisable to instruct a professional property adviser, surveyor or accountant to prepare the analysis for you.

Cashflow analysis.

(b) Uncertainty and risk

As part of the economic analysis you will need to consider the issue of uncertainty. To evaluate a best-case and/or worst-case scenario without considering the uncertainty associated with it can give misleading results, especially where the NPV of one option is more certain than that of another. Where the risk associated with different options differs substantially you should prepare a separate assessment of the risk to enable a meaningful comparison.

The simplest form of risk analysis is **sensitivity analysis**. This can be undertaken quickly and easily once the DCF spreadsheet has been prepared. The two most uncertain variables within the cashflow need to be isolated and the analysis will determine the impact of variations in these variables on the NPV figure.

Example 1.2 If the NPV of outgoings under a lease were dependent on the outcome of a rent review negotiation and the outcome of a rating appeal the sensitivity analysis might look like Table 1.3.

The expected NPV is £50 123 based on the most likely rent of £7 per sq ft and the most likely rates settlement at £3.50 per sq ft. The distribution of outcomes however spreads over a large range (between £42 710 and £58 901) and the NPV is therefore sensitive to changes in the rent and rates figure.

This is a relatively unsophisticated method of assessing risk as it fails to differentiate between the different probability associated with each of the different variables. The analysis merely provides the range of likely outcomes to enable the company to determine the extent of possible variation. More sophisticated risk analysis, known as **Monte Carlo simulation,** assigns a detailed range of probabilities to the variable elements of the discounted cashflow and provides an overall risk quotient representing the level of risk associated with the option. This can then be compared with the risk assessment for other options to assist in decision making. Computer packages are available which enable this analysis to be undertaken.

Table 1.3 Sensitivity analysis – net present value of outgoings

Rates per sq ft agreed at appeal	Rent per sq ft agreed at review				
	£6	£6.5	£7	£7.5	£8
£3	£42 710	£46 100	£49 450	£52 577	£55 089
£3.5	£42 905	£47 224	£50 123	£53 178	£56 250
£4	£43 150	£48 506	£51 066	£53 782	£57 623
£4.5	£43 459	£49 720	£52 989	£54 621	£58 901

1.2.2 Accounting analysis

Please refer to Section 1.5 for a discussion of the important issues relating to accounting for property. For all transactions you should analyse, with the assistance of your accountant, lawyer or other professional where appropriate, the impact of the transaction on the profit and loss account and balance sheet both in the short and long term.

Example 1.3 A company holds a large property portfolio including a number of properties that are surplus to requirement. Local estate agents have found a prospective assignee for the lease of one of the surplus properties. The lease has 15 years to run and the passing rent is significantly above market level, having been agreed seven years ago.

The assignee is a small company started two years ago. Their first year profits are approximately equal to the annual occupancy cost of the property including rent, rates and service charge.

The assignee has agreed to take an assignment of the lease subject to payment by the company to the assignee of a large reverse premium. The reverse premium is lower than the agent had anticipated taking into account the state of the local market; he has recommended that the company proceed with the transaction. The company needs to decide whether or not to proceed on the negotiated terms.

The economic review establishes that the assignment has significant benefits in economic terms as the NPV of the cashflows under the transaction are substantially lower than those if the company holds on to the lease. The poor financial strength of the assignee means that the transaction is risky and the worst case is the full liability for the lease returning to the company despite their having made a large capital payment. A Monte Carlo analysis would show a high risk quotient.

The accounting review will need to consider the impact of the assignment on the company's accounts in the current year and in future years. The issues that will need to be considered include:

- the significant impact of the large reverse premium on the profit and loss account for the current year, weighed against the reduction in costs which will improve profits over the 15-year period;
- the potential for a contingent liability to be shown in the accounts if the assignee is unable to discharge its obligations under the lease and the lease reverts to the company;
- the impact if the lease is assigned of a write-off in the current year of the remaining fit-out cost not yet depreciated.

1.2.3 Business analysis

This element of the review aims to bring together all business aspects of the transaction that are not reflected in the economic and accounting analysis. Some business issues will be reflected in the main financial analysis but where this is not possible the business review will ensure that they are not ignored. A quantitative analysis is preferable as it can be compared easily but where this is not possible a qualitative analysis will suffice.

Example 1.4   A service company is considering two strategic corporate property options for its office portfolio:

1. maintain the existing diversified portfolio comprising a number of properties in various locations; or
2. dispose of the existing properties and consolidate staff in one or two properties to be acquired.

The economic analysis provides an NPV of the cashflows relating to the property portfolio itself, taking into account the reduction in costs relating to disposals and increased costs relating to acquisitions. The business review addresses the wider business issues, such as:

- economies to be achieved within the business resulting from a consolidated location. These include reduced travel between offices, increased staff cohesion, elimination of duplicated business service units and duplication of functions generally;
- productivity increases relating to improved working environment;
- staff relocation costs;
- staff turnover as a result of relocation.

These three elements of the evaluation can be drawn together to enable different options to be assessed. The approach can also be adapted to assist in negotiations. Financial analysis can be a very powerful tool in negotiations. Economic analysis, coupled where appropriate with accounting and business analysis and considered from the perspective of the other party, can be used to support your proposals in negotiating transactions. Going to a landlord, for example, armed with a discounted cashflow calculation showing why they should accept a particular surrender premium can often be more successful than asking them to start the bidding. If you can keep the financial analysis within your control you have a greater chance of influencing the assumptions that will determine the NPV figure achieved.

This section has touched on the subject of appraisal techniques and the use of financial analysis in decision making and negotiations. The subject is a broad one and it has only been possible to include the most brief of discussions. The advice contained here will provide some assistance in making effective property decisions but it is advisable to refer to a specialist text for a more detailed consideration of the issues.

1.3 ADVISERS AND AGENTS

Good advice is crucial to effective management of corporate property. Managing a property portfolio involves a broad range of skills and a wide knowledge base. It encompasses a number of technical areas, including law, finance, accounting and valuation. The key to undertaking each area effectively is to take advice when necessary and delegate to agents or contractors where this is cost effective.

For small occupiers the property management function may only require instruction of advisers intermittently. Large property users may retain agents and lawyers on a wide range of projects, such as portfolio and building management, acquisitions and disposals, as well as more comprehensive outsourcing of facilities management.

In all cases the decision whether to instruct an outside agent or adviser must take into account the size of the project, the nature of the work, the cost of the advice and the in-house resources available. The advice must be tailored to the size and nature of the job: it will not, for example, be appropriate to instruct a full complement of advisers on a small acquisition.

However, you should be aware of the potential false economies associated with cutting back on advice. Often the potential savings to be gained from instructing the right adviser can significantly outweigh the cost of the advice. Use careful judgement in deciding to do without a particular adviser or to place an instruction on the basis of cost not quality.

The difficulty with saving money on professional advice is that it is hard for a client to know when it is safe to do so. It is sometimes said that the best legal document is the one that stays in the drawer and never has to be referred to. Good advice tends to be less visible than bad and it is sometimes hard to see the benefit of expenditure on professional services.

Advisers can add value in a number of different ways. Usually technical knowledge and skills are crucial; in some cases market knowledge is important; sometimes advisers simply offer resources where in-house staff have reached their capacity. The property manager will probably have cause to call on the services of lawyers, surveyors, valuers and estate agents at some time. Additionally, advice on strategic issues such as management of the property function, occupancy cost management and budgeting, may require input from management consultants, accountants and economic consultants.

Being a client is not a simple task. To get the most from your advisers and agents you need leadership, trust and teambuilding skills. You should involve your advisers in the decision-making process, keep them fully informed of your objectives, targets and strategy and encourage them to take a proactive role. All too often clients treat advisers as if they were adversaries, supplying them with information on a 'need to know' basis and discouraging them from getting involved in areas outside their strict remit. This is not the best way to get the most from your advisers.

Being a client is not a simple task.

1.3.1 Choosing advisers and agents

Before instructing an adviser you should clarify explicitly your objectives and requirements. Consider the particular requirements of the job and the qualities needed in the adviser. The importance of different qualities will depend on the nature of the project, but the following considerations will often apply:

- technical knowledge and skills, such as legal knowledge, market knowledge and so on;
- supplementary knowledge and skills such as negotiating and business management;
- ability to work as part of a team and to communicate effectively with the client and other advisers;
- the size of firm – niche player, middle sized or large national firm. This is often a choice between resources and breadth of expertise on the one hand and specialist advice on the other, although some of the larger firms are able to provide both;
- fee basis.

Ideally, advisers should be chosen on the basis of previous work or a personal recommendation. Where this is not possible, much can be gleaned from the adviser's marketing information and personal approach. It is often dangerous to make cost the primary consideration. In some cases the cheapest adviser can be the best, but this is unusual as the adviser who can sustain higher than average fees in this highly competitive market often has something special to offer.

As an occupier you will want to ensure that the adviser or agent is tuned into the needs of occupiers generally and is able to understand your wider business needs as well as the specific property issues. Often you will want to obtain fee proposals from a number of firms and in some cases it may be appropriate to interview each of them. You should find out which personnel will be doing this work to ensure a partner or director does not simply act as a figurehead and delegate the work to juniors.

A first port of call in finding an adviser can be the relevant professional body. These include the Royal Institution of Chartered Surveyors (RICS), the Incorporated Society of Valuers and Auctioneers (ISVA), the Royal Institute of British Architects (RIBA), the Law Society and the Institute of Chartered Accountants (ICA). The contact details for these and other professional bodies are listed in Appendix C. You may also wish to consider whether the firm is registered as having met the Quality Assurance standards of British Standard 5750. This simply confirms that the firm has established procedures and systems to ensure that client needs are met effectively.

1.3.2 Instructing advisers and agents

The quality of advice can be dependent on the quality of instructions. All too often clients are themselves unclear about exactly what they want from the adviser. The result can be an unhappy client who blames the adviser for a poor job. It is important that you are clear about what you expect to achieve and that you communicate this effectively to the adviser. You should explain your wider business objectives as well as the specific requirements of the job. Your letter of instruction should set out in detail the adviser's role and the work they are expected to do. In some cases uncertainty is caused by the lack of client understanding of the nature of the job, especially in technical areas where there is a knowledge gap between the client and adviser. This problem can usually be resolved by taking time to clarify all the issues with the adviser prior to agreeing instructions.

The letter of instruction should also clearly set out the fee basis. If any issue is left unclear it may give rise to a dispute later. When deciding on a fee basis the following options should be considered:

- ad valorem fee (sometimes known as a commission, incentive, performance or contingency fee) – a fee calculated as a proportion of the value of the job, such as the sale price, rental figure or savings achieved. This basis is commonly used for agency and negotiations work and has the advantage that the size of the fee is related to the size of the job and is usually only payable if the adviser is successful. The arrangement also has the advantage of stimulating the agent or adviser to achieve the best price. However, in some cases, particularly where the value of the job is high, this fee basis can result in a disproportionately high fee in comparison with the work done. Additionally, where your concern is to negotiate the lowest possible figure, for example when agreeing a lease or a rent review, an ad valorem fee may not motivate an agent effectively to promote your interests.

- quantum meruit fee – this is literally a 'fair basis for the work done', but in practice is normally a time-based fee. The adviser can charge for any time spent on the job at agreed charge-out rates. In many cases this is subject to a cap and this is usually on the basis that the work will be completed within the capped fee. The quantum meruit fee has the advantage that advisers tend to be motivated as they know they will be paid properly for the work, although this may not be the case where the cap is low. You should satisfy yourself as far as possible that the adviser will not generate work with the aim of increasing the fee. You should ask the adviser to provide regular (at least monthly) updates on the time incurred.

- fixed fee – a lump sum fee agreed at the start of the project which does not vary according to the size of the job. You should be clear and explicit about what is and is not included within the fee. Disputes often arise where advisers quote a fee based on a particular amount of work and require a further fee when this turns out to be underestimated. Clients are often attracted to this fee basis as it gives certainty. You need to satisfy yourself that the work will be done properly if it turns out to be more than expected, as advisers can become demotivated once the fee has effectively been exhausted and they consider themselves to be working for free, unless the job is a 'loss leader' designed to bring in follow-on business.

For certain areas of work professional bodies provide recommended fee scales. These are not normally mandatory but it is worth referring to them.

The fee basis will normally be quoted by the adviser and many have standard terms. However, in many cases, especially where the instruction is attractive, the adviser may be receptive to alternative fee arrangements if you suggest them. Bear in mind though that if you increase the adviser's risk you will normally have to compensate him with a higher total fee. This is the main rationale behind ad valorem fees for disposals, where the risk of abortive work is high and the total fee has to be high to compensate. It is sometimes appropriate to combine the different fee bases, for example a fixed fee element with an element based on the value of the project.

Once the advisers are instructed, to get the most out of them you should continue to take an active role in the project. Keep in regular contact and check that they are doing the job properly. When you receive reports and advice make sure you understand fully what is said and all its implications. Do not be embarrassed to ask for explanations of technical jargon.

1.3.3 Outsourcing

Outsourcing is a special form of agency where a company's in-house services are contracted out to another firm. It is commonly used for facilities management and administration departments. The following are some of the reasons why companies choose to outsource:

● to gain access to specialist skills and knowledge;
● to disperse risk;
● to free resources, such as capital and labour and to avoid deflecting in-house resources from the core business;
● to reduce cost.

This is discussed further in Section 4.7 in the context of facilities management.

1.3.4 Professional conduct

The professional conduct of advisers and agents is governed by the common law, statute, the rules of the relevant professional body and your particular contract. The rules that apply in a particular case will depend on the contract, the nature of the work and the type of adviser. The following are some important issues which may be of relevance:

● conflicts of interest – the adviser will normally be obliged to inform the client of any conflict between the interests of the client and those of the adviser, another client or a 'connected person'. In some cases the adviser will have to decline instructions or withdraw from an instruction but normally this will be at your discretion. Sometimes a conflict between two clients of the same firm can be resolved by establishing a 'Chinese wall' where confidentiality is maintained between parts of the firm. If this arrangement is suggested you need to be satisfied that the firm has effective procedures to maintain confidentiality;

- professional indemnity insurance – the professional body or other organization regulating your adviser or agent will almost certainly require that their members hold adequate insurance to cover potential claims for negligence. You would be well advised to obtain confirmation that such insurance is in place, and to obtain details of the cover, particularly the maximum amount of any one claim, to ensure that it will be sufficient for your transaction;
- clients' money – there is detailed legislation and regulations provided by the professional bodies regarding monies held by agents and advisers on behalf of their clients. The rules concern the bank accounts in which they are held, interest, records of dealings, payments allowed and other issues.

There are a range of other rules relating to issues such as acceptance of inducements, confidentiality, and quoting fees. If you have a particular concern on any of these or a related issue, you should take legal advice and/or consult the relevant professional body.

In the majority of cases clients have a good relationship with their advisers and agents and are satisfied with the service provided. In rare cases you may feel the adviser has fallen short of the expected standard and you may wish to resort to a legal resolution. In addition to any contractual duty, the professional has a duty in tort to exercise reasonable skill and care in the exercise of his instructions. The definition of this will usually relate to the standard of expertise and skill exercised by the average member of the profession and good practices recognized by a body of opinion within the profession. The adviser is not therefore expected to demonstrate exceptional ability, but may be negligent if he falls below the accepted standards of the profession. In some unusual cases the contract between the professional and the client will provide for a greater standard of performance, such as an obligation to provide a product that is fit for its purpose. This is sometimes applied in the case of construction advice.

Bear in mind that if an action for **negligence** or **breach of contract** is to be pursued it must normally be initiated within six years of the date on which the breach occurred. However, in cases of negligence only it is possible that if you did not know of a breach of duty until more than six years from the date it occurred, you may still be able to bring proceedings calculated from the date you had knowledge of the breach. In these circumstances you must issue proceedings within three years of the knowledge and within 15 years of the original breach. The measure of damages will depend on the precise nature of the action and may relate to the costs incurred or the diminution in the value of the relevant asset.

In the event of a claim you may be reliant on the adviser's professional indemnity insurance. This insurance will, in general terms, provide cover in respect of claims in circumstances in which it is established that a professional has failed to exercise reasonable skill and care in the performance of his duties. The insurance does not cover every claim which a professional may face. In particular, cover is usually limited to professional negligence and does not apply to cases where there is a breach of contractual duty but no negligence. The insurance policies are normally annually renewable and must be in force when a claim is made.

1.4 PROPERTY VALUATION

Whether the occupier is acquiring or disposing of property, obtaining a secured loan or dealing with valuations in company accounts, property valuations will have an important role in decision making. It is important that you have an understanding of their function and an appreciation of the main elements of the valuation process. Valuation can play an important role in any situation where analysis of the market is required. This section provides an explanation of the key issues to enable you to commission and interpret valuations effectively.

Property valuations can be relevant in a wide range of circumstances. The main situations relevant to the occupier are acquiring and disposing of occupational property including leases, long leaseholds and freeholds, taking out secured property loans, preparing and reading company accounts in which property assets and liabilities are represented, and management accounting. In particular, if you are buying or selling a freehold or long leasehold property you will normally need to obtain a capital valuation, and rental valuations are often important when acquiring or disposing of space held on a lease. Other uses include investment purchases and sales, takeovers and mergers, equity financing and investment, development appraisal and financing, joint venture agreements, financing instruments such as property unit trusts, and syndicated and securitized loans.

Before turning to the valuation process, we need to establish what is meant by value and to clarify the distinction between **value** and **worth**. The worth of an asset to an individual or organization is distinct from the price at which it would be exchanged in the open market. Market value is usually driven by the worth of the asset to the individual bidders who dominate the market, and many valuation techniques analyse value by reference to this. But market value is different from the assessment of worth made by any one particular investor, and while value and worth are related, it is important the two concepts are not confused.

The second important point to bear in mind is that while valuation can play an important role in decision making, it is not a panacea; it does not eliminate the need for understanding, analysis and judgement. It is a snapshot at a point in time and does not take the place of analysis. The mistake that has been made in the past by some lenders has been to treat valuations as if they were a guarantee of future value; they are only a barometer for the state of the market at one particular point in time. Of greater use than the valuation figure itself is the commentary, analysis and advice provided by the valuer which gives an indication of how the property might be affected by changes in the market, and the long-term market value.

It is important that the occupier understands what a valuation can and cannot do, how to get the most from the process and how to interpret effectively valuations prepared for you or for others. In order to get the most from the process you need:

- to understand how to commission a valuation including choosing and instructing the valuer, the basis of valuation to be adopted, and interpreting the report;
- a broad appreciation of the methods used by valuers to prepare rental and capital valuations.

The following sections provide an overview of these key issues with the aim of promoting an intelligent informed approach to commissioning and interpreting valuations. The framework has had to be simplified, but an effort has been made to achieve the level of detail necessary to cover all the important issues relevant to an occupier. Specialist valuations for rating, insurance and compulsory purchase are discussed in Chapters 1, 4 and 5.

1.4.1 Commissioning valuations

The effectiveness of the valuation process is dependent on the quality of the instructions and interpretation of the report as much as the performance of the valuer. The most important elements in the process are:

- the choice of valuer
- instructions and terms of engagement
- the basis of valuation and additional assumptions
- the valuation report.

The issues should be discussed with the valuer, and where other professionals with some expertise in the field are available, such as property managers, accountants and lawyers, it may be helpful to consult these.

The area is complex and where you do not have training or experience in dealing with valuations it can sometimes be helpful to instruct an adviser to help manage the process. This may not be practical for small one-off projects but for larger projects, such large acquisitions or sales and portfolio valuations it is often cost effective. The advantage is that the adviser will have both the skills and the time to ensure that the maximum benefit is secured from the process. The adviser should normally be a valuer or consultant with valuation expertise who has the skills necessary to understand your business needs and to ensure that the objectives of the valuation are achieved effectively.

Valuation practice is regulated by the Royal Institution of Chartered Surveyors (RICS), the Incorporated Society of Valuers (ISVA) and the Institute of Revenues, Rating and Valuation (IRRV) and the rules relating to the choice of valuer, basis of valuation, terms of engagement and valuation reports are contained in the **Red Book** (RICS, 1995). The *Red Book* applies to all valuations carried out by members of these organizations, except for some special exclusions, such as valuations undertaken in the course of pure estate agency work and those for rating and compensation. For valuations where the *Red Book* rules are not mandatory, they are best practice to be applied in most cases. The book contains useful information relating to the valuation process, the basis of valuation and valuation reports. Ideally, before commissioning a valuation, you should obtain a copy from the RICS and familiarize yourself broadly with the contents.

(a) The choice of valuer

The principles for instructing any adviser apply equally to appointing a valuer (refer to Section 1.3 for a detailed discussion). Rules and guidance are contained in the *Red Book* which defines types of valuer according to their training, experience, professional qualifications and degree of independence. Special statutory regulations also apply to the choice of valuer for certain specialist valuations, such as those carried out under Stock Exchange rules or for property unit trusts. In general, before instructing a valuer you should confirm that:

- the valuer is a professionally qualified member of the RICS, ISVA or the IRRV. These organizations provide lists of appropriately qualified members. (See Appendix C for their contact details.) In some situations specialist valuers may be needed, such as an accountant or non-qualified market expert, usually working in conjunction with a qualified valuer. The aim should be to instruct the most appropriate valuer for the job;
- the valuer has the relevant experience and competence in the valuation of the type of property concerned and sufficient knowledge of the local market. It is often advisable to instruct a locally based valuer, but in some cases an independent and objective view is needed and a non-local valuer is more suitable. The client should consider the objectives of the valuation and the best means of achieving these;
- the valuer has no conflict of interest, such as a relationship with an interested party;
- the valuer has adequate professional indemnity insurance cover;
- the fee basis is appropriate for the work and level of expertise involved.

Fee levels have fallen significantly in recent years. **Fees** for straightforward capital valuation work usually fall within the range 0.125–1%, but fees as low as 0.01% have been known. Where a very low fee is quoted you should consider carefully whether the service will be compromised as a result. If the fees quoted are low enough it can sometimes be worth obtaining two separate valuations for the same property, although such a course of action should be considered carefully as it could affect your ability to pursue a negligence action if the valuation turns out to be inaccurate.

There has been debate recently about the potential **conflict of interest** where valuations are undertaken by firms dominated by an agency practice, where the independence of the valuation can be compromised. The valuation fees in these cases can be low as the valuation work is expected to bring in agency instructions. The client should consider carefully whether the independence of the valuation may be in doubt, and in some cases it may be appropriate to commission a separate valuer to undertake the work.

(b) Instructions and terms of engagement

You will need to provide the valuer with the following information:

* full details on the subject property including any identification plans available;

The valuer should have no conflict of interest.

- details of the legal interest together with copies of all documentation such as leases, licences, rent review memoranda, planning consents;
- any relevant reports or information, such as building surveys, earlier valuation reports, planning documents and local searches.

The terms of engagement should always be in writing and should clearly set out the purpose of the valuation and any relevant background information. The valuer will normally provide the suggested terms, normally the model terms provided by the RICS, but standard terms may need to be varied. The important issues that need to be addressed prior to appointment include:

- the valuation basis (see below);
- the assumptions to be adopted in relation to contamination, deleterious materials, structural defects and wants of repair. In some cases a separate building survey or contamination survey may be required before the valuation can be completed. Where there is any possibility of contamination the valuer should make enquiries of the local authority and other relevant authority, such as the Environment Agency;
- any other assumptions, such as any restricted marketing period or legal issues. In some cases, the valuer may need to reconsider his valuation if further information shows assumptions to have been inaccurate. Artificial assumptions about the property or the legal interest held should be avoided wherever possible as they may diminish the validity of the valuation;
- the fee basis.

(c) Basis of valuation

The market value of an interest in property is normally the price at which it would sell on the open market at a point in time. This may appear to be a relatively simple concept which should not involve significant complication. Where an asset is uniform and is traded regularly, such as equities and bonds, its price can be established very simply from market data. In contrast, the valuation of a property interest usually requires a set of artificial assumptions and sometimes relatively complex analysis to enable a figure to be produced.

Properties differ in location, age, style, use, condition and a range of other characteristics. The skill of the valuer is needed to establish the price for which one property would sell based on the prices for which other properties have sold previously. Not only are the comparable sales usually historic but they relate to properties which differ, to a greater or lesser degree, in terms of age, location and so on.

Before the valuer can get on with the job, he needs to establish and agree with the client what is meant by value in the particular case concerned. The appropriate definition of value to be used depends on the purpose of the valuation. The aim must be to choose the definition that achieves the most fair and reasonable view of value having regard to the purpose of the valuation, the client's wishes and the valuer's professional duty to any potential wider audience.

The *Red Book* provides a range of different bases for rental and capital valuations, each of which is suitable in certain situations. You should consider which is the most appropriate basis taking into account the purpose of the valuation and the nature of the property. You may wish to obtain a valuation on one or more bases. The valuer and any other advisers instructed can assist you in the decision.

In addition to special definitions relating to valuations of plant and machinery and other specialized valuations, the *Red Book* contains the following valuation bases.

Open Market Value (OMV)

This is the commonly used basis for valuing investment property and is defined as: 'the best price at which the sale of an interest in property might reasonably be expected to have been completed unconditionally for cash consideration on the date of valuation assuming:

- a willing seller. [This is necessary to replicate a market situation.];
- that prior to the date of valuation, there had been a reasonable period (having regard to the nature of the property and the state of the market) for the proper marketing of the interest, for the agreement of price and terms, and for the completion of the sale. [The marketing period is assumed to have elapsed prior to the date of valuation. This has been at the centre of much debate recently resulting in a new definition (Estimated Realization Price), discussed below, in which the marketing period is assumed to elapse after the valuation date resulting in a need to determine the value that will apply at the end of the marketing period.];
- that the state of the market, level of values and other circumstances were, on any earlier assumed date of exchange of contracts, the same as on the date of valuation;
- that no account is taken of any additional bid by a purchaser with a special interest. [This assumption means that the Open Market Value may not be the highest price that would actually be paid in the market.];
- that both parties to the transaction had acted knowledgeably, prudently and without compulsion.'

The Open Market Value can include '**hope value**' for any alternative use or for the realization of any **marriage value** (see below) but only in so far as such value would be reflected in the open market by prospective purchasers without any special interest.

The Open Market Value is equivalent to the 'Market Value' as defined by the International Valuation Standards Committee and should produce the same valuation figure.

Open Market Rental Value (OMRV) is based on the same assumptions as OMV, with additional assumptions relating to the terms of the letting which are necessary for establishing a rental value.

Existing Use Value (EUV)

This follows the definition of Open Market Value but with additional assumptions that the property can be used for the foreseeable future only for the existing use and that, subject to certain exceptions, vacant possession is provided on completion of the sale.

The basis is used only when valuing for financial statements property which is occupied for the purpose of a business, and occupiers will find this basis is appropriate for many occupational properties. It is the **Net Current Replacement Cost**, which is the price that would have to be paid in the open market to replace the property. It ignores any alternative use of the property, any element of hope value for an alternative use, any value attributable to goodwill and any possible increase in value due to special investment or financial transactions such as sale and leaseback.

Estimated Realization Price (ERP)

This is similar to Open Market Value with the main difference that a reasonable marketing period is assumed to commence at the valuation date so that the completion of the sale is assumed to be later than the valuation date. The valuer needs to assess how long a period would be reasonably necessary to market the property properly to achieve the best price, and then to specify the assumed period. The valuation figure is the value in the open market at the end of that period. The definition has caused controversy amongst the valuation profession as a result of the need to assess market trends and determine value at the end of the marketing period.

Estimated Future Rental Value (EFRV) represents the rental value after a reasonable marketing period.

Estimated Restricted Realization Price (ERRP) and Estimated Restricted Realization Price for the Existing Use (ERRPEU) as a fully equipped operational entity valued having regard to trading potential

These are forced sale valuations. They are similar to Estimated Realization Price and Existing Use Value except that they assume that completion of the sale takes place after a limited marketing period to be agreed with the client. Again, there is a need to forecast value at a future date.

Depreciated Replacement Cost (DRC)

This is used for the valuation of specialized properties which, due to their nature, are rarely, if ever, sold in the open market to a single occupier for a continuation of their existing use. These properties include schools, hospitals, oil refineries, chemical works, power stations and any property where there would be no market for a sale to a single owner occupier for the continuation of the existing use as a result of the property's geographical area, construction, arrangement or size.

The Depreciated Replacement Cost is the cost of replacing the land and buildings, depreciated to reflect age, condition and obsolescence.

(d) The valuation report

The *Red Book* sets out the information to be included in the report. In addition to details such as the valuation date, the basis of valuation, assumptions and descriptions, the report should contain details of the methods of valuation and comparable market transactions and an assessment of long-term performance and market sensitivity. There should be a full commentary on all issues which have implications for current or future value, such as the design and state of repair, use, age, location, market conditions and trends, lease terms and tenant covenants.

The commentary is the most important element and should be your focus when analysing the report. It is this section that provides the basis on which to evaluate the long-term value of the property. You need to understand the way market changes are likely to affect the value of the property over the life of your interest taking into account its attributes. This analysis will assist you in assessing the degree of risk of falls in value and the prospects for rental and capital growth. If you are taking a lease, these prospects will affect the rent you will pay at review and lease renewal; if you are purchasing a long leasehold or freehold interest, the value of your capital is at stake.

You should try to get behind the valuation figures provided and you should not hesitate to ask for explanations from the valuer and comparables to support the figure. You should ensure that you broadly understand how he has arrived at the valuation figure.

The fall in the property market in the late 1980s and early 1990s led to a spate of **negligence** actions against valuers following default by borrowers on property loans. A valuer may be liable for losses where it is shown that he did not exercise the skill and care of a reasonably competent professional. There will normally be an acceptable margin of error which will depend on the particular circumstances, in one case determined at 10%. Once negligence is established the valuer may be liable for the whole of the losses incurred by the client as a result of having made the loan or purchased the property. This will not include the fall in value resulting from a change in market conditions since the valuation. Valuers' negligence is a complex issue; if you consider you have suffered a loss as a result of a valuer's negligence, legal advice should be taken.

1.4.2 Valuation methods

(a) Rental valuations

The rental value of a property is the annual rent that would be paid in the open market reflecting the valuation basis adopted. The occupier may encounter rental valuations when acquiring or disposing of a lease and they are used as the starting point for some kinds of capital valuations (discussed below). Valuations for rent review and rating are undertaken on different bases and are discussed separately in Sections 4.4 and 1.6.4 respectively.

The process comprises the analysis of **comparable evidence** to establish the rent a tenant would be prepared to pay for a lease on the property having regard to its attributes, such as location, facilities, design, state of repair, and the terms of the lease. The latter is crucial – issues such as the length of lease, rent review provisions, repair and service charge obligations and other lease terms can have as significant an impact on value as the physical attributes of the property. The valuer analyses the comparable transactions to arrive at an underlying rent per square metre or square foot and adjusts this figure to reflect differences between the comparable and subject property.

When analysing comparable transactions the valuer has to take into account any **incentives** included in the rental package and devalue these to arrive at a net effective rent. It has become customary to agree a headline or 'face' rent together with concessions which reduce the effective rent (Section 3.4.5). Landlords have preferred this type of structure as it enables them to disguise the essence of the transaction and also increases the rental figure used as the base level at the upwards only rent review. The incentives are usually rent-free periods or capital contributions but can involve more complex arrangements like takeover of existing properties.

The best method for analysing incentives is usually discounted cashflow which can accommodate both straightforward and complex rental concessions. There is disagreement about the period over which incentives should be analysed. Some argue for analysing them only in relation to the period up to the first review as it is argued that the concession is part of the initial rental package only. Others argue for analysis over the full length of the term. The period should reflect the circumstances of the transaction and the period during which the agreed rent applies. Usually the incentive relates to the headline rent agreed up to review and the analysis should therefore be over this period. Where the market level is not expected to rise above the face rent before the first upwards only rent review, it may be appropriate to analyse the rent over the period until the second review or to make an assessment of the risk of the higher rent applying.

In some cases it is necessary to exclude from the analysis any concession that aims to compensate the tenant for the cost of fitting out and the fitting-out period during which the tenant cannot occupy, but this will normally only apply where the landlord receives some benefit from the works and they are not therefore a pure inducement. It also depends on the basis on which the subject property is to be valued: if the property to be valued is fitted out then the rent-free period may need to be taken into account to ensure the comparable is analysed on an equivalent basis.

For simple transactions involving only rent-free periods or capital contributions the effective rent can be calculated quickly using a spreadsheet computer package to calculate the present value of the cashflow as shown in Tables 1.4 and 1.5. In the examples shown the effective rent is the calculated by capitalizing the income flow and then decapitalizing the present value figure achieved using the same discount rate. In some cases the capitalized figure (present value) would be decapitalized on a straight line basis without reflecting the timing of the payments as has been done here. Where this is the method commonly adopted in the market concerned it may be the most appropriate.

The effective rent shown is the equivalent rent that would be payable in each year of the term with no rent-free period. If this rent were payable the present value of all the payments over the term would be the present value shown in the tables.

As part of the analysis of comparable transactions an adjustment needs to be made to reflect any changes in the market between the date of completion of the comparable transaction and the valuation date for the subject property. Where the market is changing rapidly or where there are very few recent similar transactions the valuer's judgement of the market plays a significant role and there is greater potential for error.

In straightforward cases, the valuation will comprise a simple three-step process:

- an assessment of the appropriate rent per square foot from comparable transactions;
- a calculation of the total rent based on the rent per square foot and the floor area (refer to Appendix A for a note on the standard definitions used for calculating floor areas);
- any relevant end allowances, such as a proportional discount for poor layout (often about 10%) or poor natural light, or a proportional increase for a return frontage in retail premises (often 10–20%).

For office space it is common to apply different rates and allowances for different parts of the building, often on a floor by floor basis, for example where upper floors enjoy better light and views.

For retail properties a system of **zoning** is generally used to enable a valid comparison between different properties. The value of retail space relates to its proximity to the ground-floor shopfront which is the most productive sales space. The value of the space is therefore analysed as a proportion of the most valuable area closest to the ground floor shopfront (called zone A) with space in the second zone valued usually at half the zone A rate, in the third zone at one-quarter and so on.

Other parts of the shop are valued at a rate reflecting their position. Basement and first floor retail space, for example, is often valued between A/8 and A/12. The ground floor zones are usually 20 ft (6.1 metres) deep, but 30 ft (9.1 metres) zones are sometimes used, primarily for valuing shops in London's Regent Street and Bond Street. Figure 1.5 shows an example of the zoning of a shop.

Table 1.4 Effective rent analysis – calculation over ten-year period

Face rent: £28.50 per sq ft
Rent-free period: 1 year
Term: 10 years
Discount rate: 9%
Effective rent: £24.43 per sq ft

	Present value	Year 1	Year 2	Year 3	Year 4	Year 5	Year 6	Year 7	Year 8	Year 9	Year 10	Total
Annual equivalent (years purchase, 10 years at 9%)		–	£28.50	£28.50	£28.50	£28.50	£28.50	£28.50	£28.50	£28.50	£28.50	£256.50
£24.43	£156.76											

Table 1.5 Effective rent analysis – calculation over five-year period

Face rent:	£28.50 per sq ft
Rent-free period:	1 year
Term:	5 years
Discount rate:	9%
Effective rent:	£21.78 per sq ft

Annual equivalent	Present value	Year 1	Year 2	Year 3	Year 4	Year 5	Total
£21.78	£84.71	–	£28.50	£28.50	£28.50	£28.50	£114.00

The main principle observed is to value using the same basis as the analysis; if the comparable transactions are analysed using 20-foot zones the subject premises should also use these zones. This ensures that a consistent rate is applied and the comparable evidence is not distorted.

The zoning system is currently widely used for most retail properties, although large properties and department stores are often valued wholly or partly on an overall basis without any zoning. As shop layouts change and the differential between front and rear space diminishes, the zoning system may become less widely used.

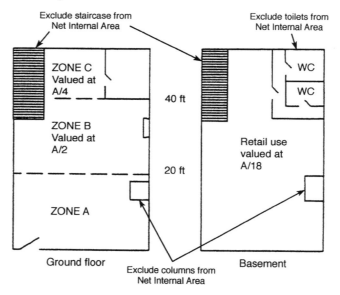

Figure 1.5 Retail valuation using zoning.

(b) Capital valuations

The techniques for valuing property interests are relatively simple in principle but complex in practice. This section provides an introduction to the subject but for a more detailed explanation, Britton, Davies and Johnson (1980) and Enever and Isaac (1994) provide a good starting point (see Further Reading section).

One word of warning when reading this section and if you find yourself dealing with valuation techniques: the terminology used can be confusingly variable. For example, the equated yield is sometimes referred to as the equivalent yield, and the cashflow method is often called a discounted cashflow (which fails to differentiate it from its growth explicit counterpart). The terms used here are those most commonly adopted but you will need to be alert to alternative uses.

The four main valuation methods are illustrated in Figure 1.6:

- direct capital comparison
- investment method
- profits method
- discounted replacement cost methods.

Direct capital comparison is the simplest and most direct approach and comprises a comparison of capital values. It is only appropriate where the attributes of different properties are similar and where value can be established directly from comparable evidence taking into account the size, location and

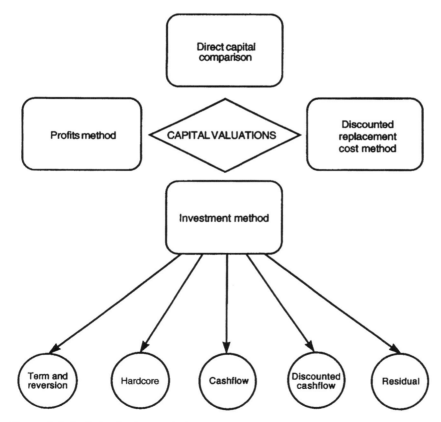

Figure 1.6 Capital valuation methods.

Table 1.6 Investment method of valuation – applications of the five approaches

Approach	Description	Application
Term and reversion	Term income and reversion income valued separately at different yields to reflect different risk, growth etc.	Especially investments with fluctuating income, e.g. over-rented property
Hardcore	Core income and top-slice income valued separately at different yields to reflect different risk, growth etc.	Especially investments with a steady income stream, e.g. prime reversionary investments
Cashflow	All income valued at the same yield reflecting risk, growth etc.	Especially investments with fluctuating income
Discounted cashflow (DCF)	All income valued at the same yield reflecting risk but not growth (growth reflected in the assumed income profile instead)	A growth explicit alternative to the other methods
Residual	Development costs deducted from projected value after redevelopment	Properties with development potential

other attributes of the property. It is the first choice valuation method but it is rarely appropriate as it cannot be used where the value relates to complex property specific attributes, such as the income flow that can be generated. In general, it is used for valuations in markets dominated by owner occupation, where analysis of the return to an investor is not relevant. It is commonly used for residential valuations and for owner-occupied industrial properties in certain areas where owner occupation is common.

The **investment method** is the most widely used method for commercial property where the market is dominated by properties held as investments and let to tenants on occupational leases. The process comprises capitalization of the projected income stream from the property reflecting any contracted rent where the property is let and the estimated rental value. The principle is to establish the price that would be paid by the highest rational bidder in the market taking into account the assumptions included in the basis of valuation chosen. It is assumed that the most any rational bidder will be prepared to pay is the total value of the income stream expected to be generated from the property. The value may not be the highest price that would be paid in the market if for example a special purchaser is to be ignored, such as for Open Market Value.

There are several techniques used for this process (the five main ones are listed below) and much debate about which is the most appropriate. The rule is to follow the market. The property should be valued using the method which dominates the market concerned. This will depend on the type of investor in the market and their objectives, but in many cases it is based on conventions that have developed historically rather than on any purely technical rationale. Table

1.6 summarizes the applications of the different methods but in practice the method chosen will depend primarily on market practice.

1. Term and reversion approach. The method is well suited to inclusion of void periods and a reduction in income in the reversion, so it is widely used for any property where fluctuations and interruptions to the income flow are expected.

Where the property is let to a tenant at a passing rent above or below the full rental value (FRV) or 'rack rent', the rent passing is valued for the term until the next rent review or lease termination, at which stage the rent payable is assumed to return to FRV in perpetuity or until the expiry of a leasehold interest. It is common to assume a void rental period at lease expiry to cover remarketing the property.

A **yield** is chosen which reflects the risk associated with the investment, primarily the security of the income flow and of the capital, the liquidity of the capital, and the growth potential of the income and capital. From transactions of comparable properties it is possible to calculate the yield at which a subject property would sell. The yields prevailing at any point in time reflect the perceptions of investors in the market of prospects for growth and certainty of income for property in comparison with other investments. The yield reflects factors which effect the security of the income flow, such as the quality of the tenant covenant and length of lease, together with factors which affect prospects for income and capital growth, such as the location and condition of the property and market trends. Tenant **covenant strength** is often the most important factor, especially for **over-rented** property where the **overage** (the amount by which the passing rent exceeds the market rental value) is highly dependent on the ability of the tenant to pay the rent. The yield applied to prime investments will often be 1–2% below long-term gilt edged securities while poor tertiary properties may be valued at as much as 10–15% above gilts.

Using the term and reversion method the contracted income (term) and projected income (reversion) are valued separately at yields reflecting the risk and growth prospects of each. The differential between the term and reversionary yield depends on investors' perceptions of the risk and growth prospects associated with both parts of the income.

Both the term and reversionary yields are called All Risks Yields as they are used to capitalize the projected income for the whole of the period until expiry, implicitly reflecting the potential for changes in the income flow either upwards or downwards. This contrasts with an equated yield, which does not reflect growth prospects, which have to be expressed explicitly in the calculation (see below).

The calculation of capital value is undertaken by converting the yield into a **Years Purchase** (YP) multiplier which capitalizes the income based on the yield and the period for which it is to be valued. The YP can be obtained from valuation tables (e.g. Davidson, 1989). Example 1.4 illustrates the valuation method.

Example 1.5   Term and reversion valuation for property let at below market value

An office is let at a passing rent of £5000 but has a full rental value of £10 000. The lease has 10 years left to run with a rent review after five years. It is assumed that at the lease expiry the tenant vacates and there is a rental void of one year. The term income is valued at a yield of 9%, reflecting market yields for this type of property in this location. The reversion following expiry is valued at 10% as this income is considered more risky. (Figure 1.7.)

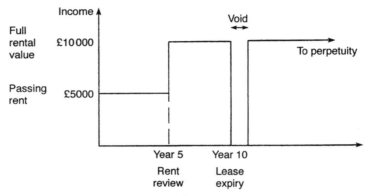

Figure 1.7 Term and reversion valuation (reversionary property).

The following is a simplified illustration of the calculation that would be undertaken:

Rent passing	£5 000	
Years purchase (YP) for 10 years @ 9% (obtained from valuation tables)	6.4177	
		£32 089
Full rental value	£10 000	
Rent passing	£ 5 000	
Rental uplift at review	£ 5 000	
YP 5 years @ 9%	3.8897	
Deferred 5 years (present value of £1 after 5 yrs @ 9%)	0.6499	
		£12 640
Reversion to full rental value	£10 000	
YP in perpetuity @ 10% deferred 11 years	3.5049	
		£35 049
		£79 778
Less costs of realization @ 2.5%		£ 1 994
		£77 784
Say		£78 000

Example 1.6 Term and reversion valuation for an over-rented property (rent passing above market value)

An office is let at a passing rent of £15 000 but has a full rental value of only £10 000. The lease has five years left to run. It is assumed that at the lease expiry the tenant vacates and there is a rental void of one year. The portion of the term income which is at full rental value is valued at a yield of 9%. The over-rented portion is valued at a multiplier reflecting a dual rate yield of 12.5% and 3% to reflect the risky nature of this income and the fact that it will discontinue at expiry and so needs to be replaced (see explanation below under hardcore valuation). The reversion following expiry is valued at 10%. (Figure 1.8.)

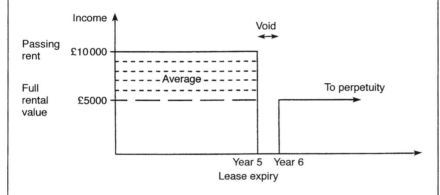

Figure 1.8 Term and reversion valuation (over-rented property).

The following is a simplified illustration of the calculation that would be undertaken:

Full rental value	£10 000	
Years purchase (YP) for 5 years @ 9%	3.8897	
		£ 38 897
Overage (passing rent above full rental value)	£ 5 000	
YP 5 years @ 12% and 3%	3.2430	
		£ 16 215
Reversion to full rental value after void	£10 000	
YP in perpetuity deferred 6 years	5.6447	
		£ 56 447
		£111 559
Less costs of realization @ 2.5%		£ 2 789
		£108 770
Say		£109 000

2. Hardcore approach. The term and reversion approach splits the income flow by time. An alternative approach is to split the income according to the level of risk associated with it. The hardcore method (sometimes called the layer method) values the core income separately from the top slice. For a **reversionary** property (let at below market value) the core income is the passing rent and the top slice is the difference between the passing rent and the full rental value; for an over-rented property the core income is the portion of income at full rental value and the top slice is the overage or additional rent receivable over the full rental value.

The hardcore method values the core income until the end of the investment, and separately values the top slice income, usually at a higher yield to reflect the higher level of risk associated with it. The method is not well suited to valuing fluctuating income, so is not generally used for over-rented properties and other properties without a steady income flow. Examples 1.6 and 1.7 illustrate the method.

Example 1.7 Hardcore valuation for property let at below market value
An office building is let at a passing rent of £5000 but has a full rental value (FRV) of £10 000. The lease has 10 years left to run with a rent review at year 5. The yield applicable to the core (passing) income is assessed at 9%. The top slice yield is assessed at 10%. (Figure 1.9.)

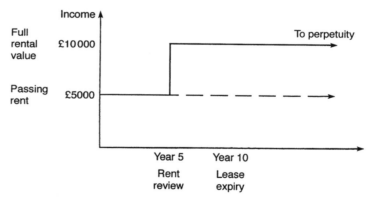

Figure 1.9 Hardcore valuation (reversionary property).

The following is a simplified illustration of the calculation that would be undertaken:

Rent passing	£5 000	
Years purchase (YP) in perpetuity @ 9%	11.1111	
		£55 556
Full rental value	£10 000	
Passing rent	£ 5 000	
Rental uplift	£ 5 000	
YP in perpetuity @ 10% deferred 5 yrs	6.2092	
		£31 046
		£86 602
Less costs of realization @ 2.5%		£ 2 165
		£84 437
Say		£84 500

Example 1.8 Hardcore valuation for an over-rented property
A shop is let at a passing rent of £15 000 but has a full rental value (FRV) of only £10 000. The lease has 15 years left to run with rent reviews at years 5 and 10.

It has been ascertained from market information that investors are assuming projected rental growth of 4.5% per annum giving a full rental value after five years of £12 500. This projected growth rate is reflected in the yield applied to the core (market) income of 10%. The top-slice income (overage) is considered to be risky and is valued at a yield of 13%.

The 10% yield applied to the core income reflects an implied rental growth rate of 4.5%. This means that a £2500 uplift in rent is already reflected in the valuation of the core income for the five years following the first review. This £2500 has to be deducted from the top-slice income for the second five years of the term to avoid a double counting of this income.

After the second review, rental growth means that the core income has risen above the passing rent and the top slice is no longer valued. The multiplier (YP) used to value the top-slice income has to reflect the fact that this income will stop after the second review. Investors expect to be able to redeem their capital at the end of the life of the investment and the top-slice income must therefore be valued at a rate that reflects replacement of the income once it discontinues. This is achieved by means of a dual rate multiplier reflecting a sinking fund rate of 3% which will replicate the capital value of the income when it finishes at year 10. (Figure 1.10.)

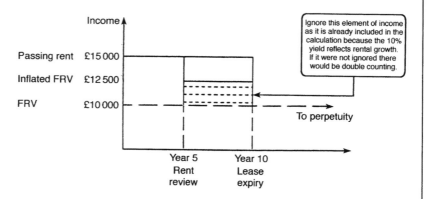

Figure 1.10 Hardcore valuation (over-rented property).

The following is a simplified illustration of the calculation that would be undertaken:

Full Rental Value (core income)	£10 000	
Years purchase (YP) in perpetuity @ 10%	10.00	
		£ 100 000
Overage (top slice)	£ 5 000	
YP 5 years @ 13% & 3%	3.1412	
		£ 15 706
Top slice after 5 years	£ 2 500	
YP 5 years @ 13% & 3%	3.1412	
Deferred 5 years (present value of £1 after 5 yrs @ 13%)	0.5428	
		£4263
		£ 119 969
Less costs of realization @ 2.5%		£2999
		£ 116 970
Say		£ 117 000

The method is not well suited to the treatment of void periods as the core income is valued throughout the life of the investment. It is common in the current market to include void periods on lease expiry and the hardcore method is therefore not widely used.

3. Cashflow approach. This approach applies a single yield to all the income as shown in Figure 1.11. The yield applied is the equivalent yield, which is the internal rate of return of the investment and reflects the level of risk associated with all parts of the income. The method is ideally suited to valuing changing levels of income over the life of the investment as the income is analysed year by year and not grouped together as in the more traditional methods. It is suited to the treatment of void periods and negative income, whereas the traditional methods are less effective at valuing these types of income flow.

The method is often used for valuing shopping centres as value is related to a fluctuating stream of income.

4. Discounted cashflow approach (DCF). The cashflow technique uses an all risks yield which reflects implicitly a set of assumptions about the risk and growth prospects of the investment. The discounted cashflow method analyses all assumptions explicitly and uses an equated yield (target rate of return) to discount the income flow to present value. The yield reflects risk but not growth and the method differs from the cashflow approach which uses a yield reflecting both risk and growth. In this model, the growth assumptions are expressed explicitly through the estimates of the income receivable each year during the life of the investment. These take into account growth at an estimated level and changes in income levels due to rent reviews, lease expiries, break options and so on.

This method is effective as it enables an explicit analysis of all income flows. It is not however widely used in the market by investors or their advis-

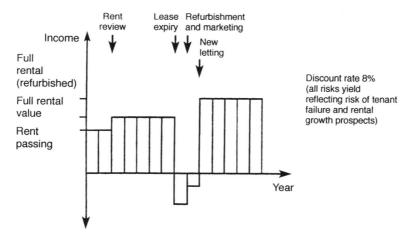

Figure 1.11 Cashflow valuation.

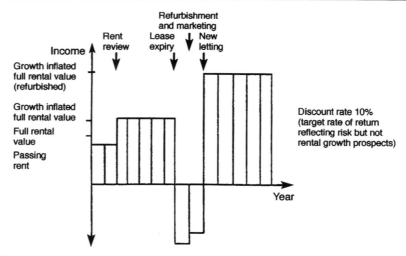

Figure 1.12 Discounted cashflow valuation.

ers for analysing investment value (although it is more often used in establishing the worth of an investment to the particular investor concerned). The approach is complex to use in valuations because of the need to establish a market risk free discount rate from comparable evidence and to translate market perceptions accurately into an explicit anticipated income stream.

For all valuations the approach adopted should be the one most commonly used in the market concerned. The DCF technique is more widely used in the analysis of worth than value, and in Section 1.2 its use in decision making is discussed in detail. In this type of analysis there is no need to establish market perceptions and discount rates, as the analysis focuses on the individual investor's perspective and does not attempt to establish market value.

The problems of valuing over-rented property have prompted a number of techniques to be developed which aim to eliminate the double counting of some of the top-slice income. These use the DCF model but analyse only the short-term income on this basis. These variants on the more traditional methods are generally known as shortcut DCF or Real Value approaches.

5. Residual approach. Where properties have development potential or potential for an alternative use this inherent value needs to be addressed in the valuation. The residual method values the property following the redevelopment or change of use and deducts the cost of development.

The value on redevelopment is estimated using one of the other valuation methods, often the investment method, in which case the income stream anticipated from the property after development is valued. Included in the redevelopment cost is the cost of finance which reflects the time taken to complete the development, and the developer's profit. Alternatively, in some cases the funding cost is excluded but income and expenditure are discounted to arrive at their present value. The residual method is the conventional technique used for valuing development land and for alternative use valuations.

The problem with this technique is its sensitivity to small changes in assumptions. This is because the value is related to the difference between end value after development and the cost of development, which is very sensitive to changes in assumptions.

A third approach to valuation is the **profits method**. For some specialized properties the income generated is highly dependent on the particular circumstances, particularly the effect of location and other attributes on the income that can be generated from the property. As a result the income cannot be estimated from comparable properties. In these cases it is necessary to look at the potential profits that can be generated from the subject property. This method is generally used for the valuation of hotels, petrol stations, leisure facilities such as golf clubs, cinemas and health clubs, and some other unusual business properties.

The profits expected to be generated from the property are capitalized either using a multiplier identified from market evidence or using a cashflow or discounted cashflow method. The cashflow or DCF methods are becoming the preferred technique and are recommended by the British Association of Hotel Accountants (BAHA) for the valuation of hotels.

Where a type of property is not regularly transacted in the market and a profits basis would not fairly reflect value, the only basis on which value can be established is by reference to the cost of replacing the building. This is termed the **discounted replacement cost method** (DRC). The properties valued on this basis are specialized buildings such as churches, schools, police stations and some industrial buildings.

The replacement cost is derived from the market value of the site and the construction cost for an equivalent building. The valuer then has to make deductions to allow for the quality of the property as existing, and in particular in relation to economic and functional obsolescence, including the cost of future maintenance of the existing building and its suitability for the present use and other uses.

The problem with this method is that it provides an analysis of cost not value. It is usually only used as a method of last resort for market valuations. A form of replacement cost valuation is used for insurance valuations without allowances for obsolescence (Section 4.4.2).

(c) Leasehold properties

The value of a leasehold interest, or premium that would be payable in the market for the lease, usually relates to the **profit rent** which is the difference between the contracted rent and the estimated rental value of the premises. Where the rent passing is below the full rental value, the lease may have a value and a premium may have to be paid to acquire it. Conversely, where the property is over-rented a reverse premium would normally be payable by the assignor to the assignee, unless the period left on the lease is short and there is no market for the interest. In some cases key money is payable by an ingoing tenant and this capital sum will be reflected in the leasehold value.

The premium payable for a reversionary property (rent below market) relates to the profit rent until review capitalized at a market all risks yield. There may be no value in the lease after the review date if the property returns to full rental value. Similarly, a profit rent does not necessarily result in leasehold value as it can be outweighed by other costs, such as dilapidations and reinstatement, and where the period before expiry is short there may be no market for the interest.

Conversely, there may be some scope for a profit rent after review if the review is not to full market value because of the items to be disregarded in valuing for rent review, particularly tenants' improvements, or where the implicit growth assumption reflected in the all risks yield includes potential for a profit rent at later stages in the term.

The all risks yield for a leasehold interest is generally higher than a freehold yield to reflect the inflexibilities associated with holding a leasehold interest and the lower growth potential, especially for short leasehold interests. The investment valuation methods outlined above can be used but it is conventional to reflect in the capitalization rate an allowance for a **sinking fund** to replace the value of the lease on expiry to compensate for the wasting nature of the leasehold (as described in Examples 1.5 and 1.7 above). The cashflow method avoids the need to use a sinking fund and this is being used more commonly as a valuation method for leaseholds.

(d) Marriage value

As a result of the difference in the mechanics of leasehold and freehold valuations, especially the yields applied and the use of sinking funds in the valuation of leasehold interests, the value of the merged interest, achieved by surrender of the lease, can be greater than the sum of the values of the two separate interests. This difference is the marriage value.

Example 1.9 An industrial unit is occupied by a tenant on a lease expiring in seven years time at a passing rent of £5000 without review. The full rental value of the unit is £10 000.

The marriage value is calculated as follows:
Value of landlord's existing interest:

Term rent	£ 5 000	
YP 7 years at 9%	5.033	
		£25 165
Reversion to full rental value	£10 000	
YP in perpetuity deferred 7 years at 10%	5.1316	
		£51 316
Value of landlord's existing interest		£ 76 481

Value of tenant's existing interest:

Full rental value	£10 000	
Rent passing	£ 5 000	
Profit rent	£ 5 000	
YP 7 years at 11% and 3%	4.1579	
Value of tenant's existing interest		£ 20 790

Value of merged interest:

Full rental value	£10 000	
YP in perpetuity at 9%	11.1111	
		£111 111

The marriage value is the difference between the total value of the two existing interests and the value of the merged interest, i.e. £111 111 – (£76 481 + £20 790) = £13 840. The landlord and tenant can each try by negotiation to secure as much as possible of this marriage value as part of the surrender negotiation.

The *Red Book* prohibits inclusion of any marriage value in valuations except in so far as any 'hope value' for a merger of interests is reflected in the market value. Marriage value can be included in some other types of valuations, for example valuations for capital gains tax assessments and valuations for disposals.

Occupiers holding long leases at rents below the market level should consider the potential for securing marriage value through a surrender of their lease, particularly if vacating the property would fit in with business needs.

(e) Glossary of yields

The **all risks yield** is any of the yields discussed below which take into account all the attributes of the investment including risks and expectations for growth (i.e. all those listed here except for the equated yield). These all risks yields can be contrasted to the equated yield which excludes growth.

The **equated yield** is the target rate of return or growth free yield. It is used in valuations in which growth is expressed explicitly in the assumed net income from the property, and can be contrasted to the all risks yield which is growth-implicit.

The **equivalent yield** is the internal rate of return of the investment. It reflects the income receivable over the life of the investment and reflects the timing of the different income flows and yield applied to these. It is the discount rate which needs to be applied to the flow of income to arrive at the capital value. It takes into account the yields applied to the term yield and the reversionary yield in a term and reversion valuation, or hardcore and top slice yields in a hardcore valuation. Example: Where a property has a rent passing of £100 000 for five years valued at 9% and a reversion to full rental value of

£150, 000 for five years valued at 10%, the capital value is £758, 500 and the equivalent yield is 9.26%.

The **hardcore yield** is the yield applied to the core income in a hardcore valuation.

The **initial yield** is the return on capital represented by the initial income and is the proportion of capital value represented by the initial rental income. Example: The initial yield on a property selling at £1 000 000 with an initial passing rent of £100 000 is 10%.

The **reversionary yield** is the yield applied to the income stream anticipated after the term in a term and reversion valuation. Example: Where the passing rent is valued at 9% and the reversion is valued at 10%, the reversionary yield is 10%.

The **term yield** is the yield applied to the income receivable during the term in a term and reversion valuation. Example: Where the passing rent is valued at 9% and the reversion is valued at 10%, the term yield is 9%.

The **top-slice yield** is the yield applied to the top-slice income in a hardcore valuation. It is the yield applied to the income which represents the difference between the rent passing and full rental value.

Note: The terms 'equated yield' and 'equivalent yield' are sometimes given different meanings than shown here. You should clarify the sense in which the term is used whenever you encounter it.

1.5 PROPERTY ACCOUNTING

Corporate property management and decision making is generally directed towards minimizing cost and maximizing efficiency, with the aim ultimately of maximizing profit in the short and long run. It is important that the property manager has regard to the accounting implications of their management decisions and the accounting impact of every transaction needs to be considered as part of the decision making process, as discussed in Section 1.2. This should include the impact both on the **balance sheet** and on the **profit and loss account**. This section provides a brief explanation of the important accounting issues of relevance to occupiers.

Accounting for properties must have regard to the following two sources of rules:

- legal requirements contained in the Companies Act 1985;
- professional rules contained in accounting standards prepared by the UK Accounting Standards Board (ASB) or its predecessor, the Accounting Standards Committee. Notable examples of relevance to property include the following: *Statement of Standard Accounting Practice (SSAP) 12 –*

'Accounting for depreciation'; Statement of Standard Accounting Practice (SSAP) 19 – 'Accounting for investment properties'; Statement of Standard Accounting Practice (SSAP) 21 – 'Accounting for leases and hire purchase contracts'; Financial Reporting Standard (FRS) 5 – 'Reporting the substance of transactions'.

Consider the impact of every decision on both the balance sheet and the profit and loss account.

1.5.1 Ownership issues – accounting for assets

If you own any freehold or long leasehold interests in property, either for occupational use or for investment, you will need to account for these assets in the balance sheet. This section highlights the main issues of relevance.

(a) Asset value

The Companies Act 1985 provides for two bases of accounting for tangible fixed assets, including properties. They can be accounted for using a pure historical cost convention, or on a modified historical cost convention under which the carrying value (the value shown in the accounts) of some assets is adjusted to market value on the basis of revaluations.

The key requirements of the **historical cost accounting** rules are as follows:

● assets should initially be recorded at purchase price or production cost;

- assets with a limited useful economic life should be systematically depreciated down to their estimated residual value over that estimated life;
- any asset which at any time is shown to have suffered a permanent diminution in value below its carrying amount must be written down immediately to its recoverable amount, whether or not its life is limited.

The directors of a company may choose to adopt the **modified historical cost accounting** rules for particular assets. The key requirements of these are as follows:

- tangible fixed assets may be included either at a market value determined as at the date of their last valuation, or at their current cost;
- where a valuation basis is used, the value replaces cost as the starting point for applying the depreciation rules;
- differences arising on revaluation are credited (for increases) or debited (for decreases) to a revaluation reserve.

Current practice demonstrates the ability of UK companies to find loopholes within the legal requirements. For example:

- companies are able to revalue some, but not all, properties. They can therefore select the most attractive properties for valuation. For example, a retail group may revalue office sites, but not retail and distribution facilities;
- although the Act refers to permanent diminutions in value, it does not make any reference to temporary diminutions in value in either the historical cost or alternative accounting rules. Therefore, no adjustment is necessary for assets where their value is beneath balance sheet carrying value provided that there is some evidence that the situation is not permanent. It is common to leave such assets at cost or latest upward valuation;
- under the alternative accounting rules, it may be assumed that whenever a change in market value occurs, it should be reflected in the balance sheet. In practice, advantage is taken of the Act's requirement to include assets at a market value determined as at the 'date of their last valuation'. Commonly, companies undertake a full valuation every three years. However, in the absence of a mandatory requirement to revalue at any set intervals, companies will sometimes claim that property values were reviewed, but that this did not amount to a full valuation. Consequently, no revaluation is booked. Thus, although a company may adopt a policy of carrying properties at valuation under the alternative accounting rules, the latest booked revaluation may have been five or more years before the balance sheet date;
- companies have argued that a temporary fall in open market value does not necessarily mean that the carrying amount of an asset is unrecoverable. This is particularly so when the asset continues to be used profitably.

It may help to understand why company law allows valuation on an open market basis. Before 1948, assets could only be carried at historical cost. The

Companies Act 1948 (now consolidated into the 1985 Act) was the first to introduce alternative accounting rules. The context was post-war reconstruction, a rising property market and companies wishing to demonstrate balance sheet strength against which to borrow. Open market value provided an available method of establishing higher carrying value. Since falls in value were not anticipated at that time, company law did not establish robust rules for dealing with such an eventuality. From a theoretical point of view, open market value is most relevant in two situations:

- for a company holding a property which it expects to sell, it is the best estimate of value at disposal;
- where a property is to be used through to the end of its life and then replaced, it is the best estimate of replacement cost of a similar asset, and hence the amount of money that the company must retain towards replacement.

This latter idea of replacement was at the heart of the **current cost accounting** (CCA) convention, which was adopted in the UK in 1980, but dropped within a few years. Under CCA, a property would be carried at the lower of net replacement cost or recoverable amount. Net replacement cost could be established either by open market value, or by indexation with allowance for the proportion of life consumed. Net recoverable amount is the higher of value in use (economic value, normally discounted cash flows from use) or value for sale (net realizable value).

The Accounting Standards Board is wrestling with the issues of accounting convention, selective and infrequent revaluation, and methodology. The downturn in property values in the 1990s has focused attention on many of the issues. The response of some companies to lower property values, and the company law and accounting issues that arise, is to switch back to historical cost from modified historical cost (i.e. giving up revaluations).

(b) Depreciation

The principle of depreciation is derived from the Companies Act and amplified in *SSAP 12*. The key requirement is:

> provision for depreciation of fixed assets having a finite useful economic life should be made by allocating the cost (or revalued amount) less estimated residual value of the assets as fairly as possible to the periods expected to benefit from their use.

The principle of annual depreciation may be represented as follows:

Annual depreciation =
$$\frac{A - B}{C} \quad \text{where: A = cost or revalued amount; B = estimated residual value; C = useful economic life.}$$

In practice, freehold and long leasehold property is often not depreciated, on the grounds that any charge would be immaterial. This may be the case where either estimated residual value is close to book value (that is, A − B is very small), or the estimated life is very long (that is, C is very large), or both. A claim that life is infinite is generally not supportable. However, a programme of continual refurbishment may support an argument that residual value is maintained, and that life is being continually extended. Such arguments are often valid in respect of leisure properties such as hotels and public houses.

It is not acceptable to argue that, due to short-term increases in market value, depreciation can be avoided unless there is evidence that economic useful life has changed, or that the initial estimate of residual value was incorrect. If the asset's life is finite, then carrying value must be consumed as the asset is used. It is similarly not acceptable to take account of increases in estimated residual value due to general increases in prices during the asset's life.

An alternative to depreciation is to use the **renewals basis of accounting** which involves charging all maintenance costs as a substitute for annual depreciation. The asset will normally have an unlimited life, and be subject to a programme of regular survey and maintenance. The method is used for major infrastructure assets such as water mains and railways where economic useful life is genuinely continually being extended to perpetuity. It is rarely appropriate for commercial property.

(c) Investment properties

The UK is the only major western country with an accounting system that maintains a separate classification for investment property. Where a property is held for its investment potential, and not for use within the business, it is argued that open market value best reflects the value to the business, that changes to open market value best reflect performance of the asset, and that depreciation is inappropriate since there is neither any estimable useful economic life nor residual value. The definition of an investment property is contained in *SSAP 19*, and is as follows:

> an investment property is an interest in land and/or buildings:
>
> (a) in respect of which construction work and development have been completed; and
>
> (b) which is held for its investment potential, any rental income being negotiated at arm's length

The following are exceptions from the definition:

- a property which is owned and occupied by a company for its own purposes is not an investment property;
- a property let to and occupied by another group company is not an investment property for the purposes of its own accounts or the group accounts.

Classification of properties as investment properties is not always clear cut. It will depend on a judgement of whether the property is held for its investment potential. Where a property does not command a market rental (or indeed any rental at all), it may nevertheless qualify as an investment property if it can be demonstrated that it is held for its capital appreciation potential. SSAP 19 represents a departure from the depreciation rules of the Companies Act 1985, and is justified as being necessary for accounts to give a true and fair view.

Prior to an amendment to *SSAP 19*, all revaluation differences in respect of investment properties were taken directly to a separate investment **revaluation reserve**. A charge to the profit and loss account occurred only where, in aggregate, the investment revaluation reserve had a net debit balance. Following the amendment to *SSAP 19*, property values and resulting surpluses or deficits are assessed property by property: permanent falls in value are required to be shown in the profit and loss account; all temporary changes in value, whether positive or negative, are reported in the statement of recognized gains and losses, and taken to an investment revaluation reserve. Where the resulting investment revaluation reserve shows a net debit balance, no adjustment is required.

1.5.2 Leasing issues

(a) Finance and operating leases

Accounting for leases is the subject of *SSAP 21 'Accounting for leases and hire purchase contracts'*. On the basis of *SSAP 21*, where, at the inception of a lease, the net present value of minimum lease payments exceeds 90% of the asset's fair value, then the lease may be presumed to be a finance lease, with the fair value of the asset capitalized in the lessee's balance sheet matched by an equal amount of deemed borrowings. If the net present value of rentals falls under 90% of the asset's fair value, then the entire asset stays off-balance sheet and rentals are dealt with as they become payable.

However, since *FRS 5 – 'Reporting the substance of transactions'*, the 90% test is not the only method of determining the type of lease. Where the net present value of rentals is less than 90% of the asset's fair value, but it is nevertheless determined that other factors indicate that the lease is, in substance, a finance lease then *SSAP 21* should be overridden and the asset capitalized on the tenant's balance sheet. Under *FRS 5*, this would not be the entire asset, but only those rights which the lessee has acquired. For example, a lease where the net present value of rentals represented 80% of the asset's fair value could be capitalized at the present value of the rentals. Where a lease forms one part of a larger transaction, it is likely that *FRS 5* will be applied to the entire transaction.

Until 1984 one of the main motivations for occupiers adopting a finance lease structure was that this type of debt was not shown on the balance sheet. This changed with the publication of *SSAP 21*, which requires finance leases to be capitalized in the occupier's accounts. There have been suggestions that operating leases should also be shown in this way: the lease payments as a liability and the capital value as an asset.

A **sale and leaseback** may or may not result in a disposal of a property for accounting purposes. As a general rule, where a leaseback qualifies as an operating lease, the sale will be accounted for as a disposal, with any profit or loss recognized immediately (provided the sale is at arm's length – any surplus over an arm's-length price would be deferred and spread over the lease term.) Where the leaseback is a finance lease, it follows that all significant risks and rewards of ownership have been retained and no disposal has occurred for accounting purposes. In this case, the asset will continue to be recorded at its previous carrying value and proceeds received will be classified as a liability which is repaid by lease rentals. Similar principles will apply to other arrangements such as lease and leaseback. More complex variations are treated on the basis of the substance of the transaction as a whole.

Charges for operating leases reflect rent only; other costs, such as insurance would be classified separately. In accordance with the **matching principle**, rent is generally allocated to profit and loss accounting periods on a straight-line basis as this matches the other costs and the income derived from the use of the property. The exception would be where another basis would better reflect the pattern of use of the asset, for example surplus property where the whole of the rental commitment may be charged in one accounting period (see below). Where the outcome of a rent review is uncertain, it will be taken into account only when it happens; if future increases in rent or other charges are agreed in advance, the full amount based on a best estimate of actual cost would be spread on a straight-line basis. In addition to allocating the operating costs for the period concerned to the profit and loss account, there is a requirement to disclose lease payments due in the following period in a note.

Alterations to property that enhance the value of the asset or result in measurable future benefit are capitalized in the balance sheet as an addition to the cost of the asset and depreciated over the remaining useful life of the asset. For a fit out undertaken on acquisition, this may be for the full term until expiry or a break option. Repair costs that do not add value to the property are expensed as they are incurred.

(b) Incentives

The Accounting Standards Board Urgent Issues Task Force has considered incentives granted by landlords to tenants, such as rent-free periods and capital inducements or contributions to fitting-out costs. Abstract 12 produced by the Task Force requires that all incentives should normally be spread on a straight-line basis over the period of the lease, except where the directors can make a case for using a different period. The Abstract is not a full accounting standard but will normally have to be followed. Decapitalization of incentives is discussed in Section 1.4.2.

(c) Surplus property

The Urgent Issues Task Force considered requiring companies to provide for all costs related to the lease during the period of expected vacancy where the property is vacated before the end of the lease. The rationale is that once the property is not operational it becomes a liability that should be recognized. The Task Force decided not to proceed with the Abstract but the matching principle will need to be satisfied and in some cases recognition of the liability is appropriate. A change in the regulations has been mooted but is unlikely to happen for some time.

Where a lease is disposed by assignment or subletting and there is a significant risk of the lease liabilities reverting, this contingent liability needs to be disclosed in a note to the accounts (see Section 5.1.8).

1.6 PROPERTY TAXATION

This section aims to provide a framework for understanding the taxation of property in the UK, and in particular to highlight the main issues affecting commercial occupiers. The regime is simplified here, so refer to a specialist textbook for a more detailed explanation (see Further Reading). You should seek the advice of a lawyer, accountant or relevant specialist on structuring particular transactions tax efficiently and minimizing your tax liability generally.

1.6.1 Income tax, capital gains tax and corporation tax

Sole traders and most partnerships are charged income tax on their profits and gains of an income nature, and capital gains tax on their profits and gains of a capital nature. Companies are charged corporation tax on their profits and gains of both an income and capital nature.

Income tax is charged on the income of UK residents, whether it arises in the UK or abroad, subject to certain deductions. A company resident in the UK is chargeable to corporation tax in respect of its worldwide income profits and capital gains and a non-resident company carrying on a trade in the UK through a branch or agency is liable to corporation tax on the income (and capital gains on UK assets) of the branch or agency.

(a) Income

The word 'income' is not defined in the tax legislation. Instead the legislation classifies amounts received under various headings, called Schedules, and an item must come within one of these headings to be charged as income. In certain instances the legislation requires capital items to be treated as income.

The **income tax** payable by an individual, either a sole trader or member of a partnership, is calculated with reference to the sum of the amounts he receives under the income Schedules, less certain payments made and allowances and reliefs available.

Similarly, **corporation tax** on companies' income profits is calculated with reference to the sum of the amounts it receives under the income Schedules, less certain deductions.

(b) Capital

Individuals are chargeable to **capital gains tax** on the total amount of their chargeable gains, exceeding their annual exemption (£6300 for 6 April 1996 – 5 April 1997), less any allowable losses. A chargeable gain or allowable loss is made when a chargeable asset (all forms of property are chargeable unless specifically exempt) is disposed of. This includes both sales and gifts.

Gains and losses are calculated by deducting from the disposal proceeds (or in some cases, from the market value at the time of disposal) the cost of the asset when it was acquired, the **base cost**, plus incidental costs of acquisition and disposal and certain expenditure which has increased the value of the asset. Some allowance is made for any increase in the value of assets due to inflation.

A company's capital gains are chargeable to **corporation tax**. The gains are computed in the same way as for individuals, except that companies do not have an annual exemption.

(c) Tax treatment of property transactions

The treatment of any profits and gains derived from the disposal of commercial land and buildings depends on whether they are considered to be held as an investment or for the purpose of trading in land. This will be determined on the basis of a range of factors, such as the purpose of acquiring the land, the length of ownership, circumstances of sale, frequency of similar transactions and whether any work was done on the property or any income received from it.

If the land and buildings are held as an **investment**, any profits and gains derived from the disposal of the land will usually be treated as being of a capital nature; profits and gains derived from **trading** disposals of land will usually be treated as income receipts. In the past this distinction was more important as the rate at which capital receipts were taxed for individuals was lower than the rate applicable to income receipts, and for companies only a proportion of capital receipts were taxed; now the rates and proportion of gains taxed are the same.

For sole traders and partnerships, there is still an advantage to being treated as investing rather than trading as the first £6300 of an individual's chargeable gains in any one tax year are exempt and certain reliefs are available on the disposal of land held as an investment which are not available if trading in land. For a company there are advantages and disadvantages of being treated as investing rather than trading. Property held by a business for its own occupation will normally be treated as an investment asset.

Acquisitions and disposals

There is no charge to capital gains tax, income tax or corporation tax on the **acquisition** of an interest in property.

The **sale** of a freehold, or **assignment** of a leasehold, acquired for occupation and use as business premises is normally considered to be an investment **disposal** and is charged to capital gains tax or, in the case of companies, corporation tax. If, after allowing for inflation, the sale price exceeds the acquisition cost plus incidental expenses of acquisition and disposal, plus expenses which increase the property's value, then a chargeable gain will be made. If not, an allowable loss may be made, which can be set off against capital gains in the same year of assessment or future capital gains. The premium paid for the assignment of a lease which has 50 years or less left to run, is gradually written-off over time to represent the depreciating value of the lease. This will reduce the base cost of the lease used to calculate any capital gain made on the disposal of the lease.

If the interest were acquired before March 1982, an election can be made to treat it as if it was acquired at its market value in March 1982 (usually giving a higher base cost).

If you **grant** a lease this is treated as a **part disposal** for the purposes of capital gains tax and corporation tax on chargeable gains. On the grant of a lease for a term of 50 years or less, part of the premium paid for the grant is treated as capital and part is treated as additional rent. Consequently, part of the premium received by the landlord will be subject to capital gains tax, or corporation tax on chargeable gains, and part subject to income tax or corporation tax on income profits. The capital portion of the premium will be the tenant's base cost for the purposes of capital gains tax or corporation tax on chargeable gains and, as with the assignment of a lease with 50 years or less to run it is written-off over time. The portion of the premium deemed to be rent is usually deductible from the tenants income trading profits (see below).

There are anti tax avoidance provisions which make gains on property acquired or developed with the sole or main object of realizing a gain on disposal, subject to treatment on disposal as income, rather than capital.

The following reliefs are available in relation to certain disposals:

- Rollover relief. Where a business makes a chargeable gain on the disposal of land and buildings occupied for the purposes of its business and the proceeds are invested in replacement land and buildings (or another qualifying business asset) within the period commencing one year before and ending three years after the disposal, a claim can be made for the taxation of the gain to be postponed. This is achieved by reducing the base cost of the replacement asset by the amount of the gain on the asset being replaced. Relief can be claimed any number of times in respect of property continually replaced.
- Other reliefs. There is also relief available for capital gains made on disposals of assets of a business, or shares in a business, by the owner of the business on his retirement. There are special rules for the treatment of intra-group transfers.

If you are disposing of an interest in property by granting a sublease, the treatment of income (such as rent received) from the property depends on which income Schedule it falls within. Income received for the use of land will usually be charged to income or corporation tax as investment rather than trading income. The income may be of a trading nature if it is derived from something more than merely the use of the land. The distinction between trading and investment income is important because, amongst other things, on disposal a property from which investment income is derived will not qualify for rollover relief (unless a compulsory purchase) or retirement relief.

Occupation

Where property is occupied by a business carrying on a trade, expenditure on the property (rent, maintenance and so on) is deductible from income profits if it is of a revenue, rather than a capital, nature and has been incurred wholly and exclusively for the purposes of the trade. If an expense is not incurred wholly and exclusively for the purposes of a trade, but is nevertheless of an income nature, it may be deductible from a non-trading income (if there is any) derived from the use of the land or buildings. Capital expenditure may be deductible in calculating any capital gains tax or corporation tax on a company's chargeable gains. Rent paid by a business for occupation of premises for its trading purposes is usually deductible from its income trading profits.

Where a business ceases to occupy premises for trading purposes before expiry of the lease, it may continue to deduct rent from trading income, provided the premises are not put to a non-trade use. If the premises are sublet, but the rent payable by the business exceeds rent received under the sublease, the difference is deductible from trading income.

Where land is being leased from a landlord who resides outside the UK and rent is not paid to a UK agent of the landlord, the tenant may be required to deduct basic rate income tax (24%) from the rental payment made to the landlord, and pay this direct to the Inland Revenue. Otherwise, if the landlord fails to pay over to the Inland Revenue the income or corporation tax due on the rental income he receives under the lease, the tenant will be pursued for the tax.

In calculating a business' income profits for tax purposes, unlike for accounting purposes, no deduction can be made for the depreciation of capital assets. However, a business can claim certain allowances against its income profits for capital expenditure on certain assets, such as plant and machinery owned and used by the business and industrial buildings. **Capital allowances** for a reducing proportion of the cost of plant and machinery can be claimed in the year of purchase and future years until allowance has been made for the total cost of the plant and machinery. A tenant can only claim allowances for expenditure on **plant and machinery** owned by him, and cannot claim allowances for expenditure on fixtures owned by the landlord, unless they fall within special provisions of the tax legislation. In addition to equipment, industrial buildings are eligible for special allowances (**Industrial Buildings Allowances**, IBAs). These buildings include factories, properties used for manufacturing purposes and certain warehouses. Allowances for a fixed proportion of certain expenditure on industrial buildings can be claimed each year until allowance has been given for the total expenditure.

Certain areas of the UK have been designated enterprise zones by the government. Any expenditure incurred or contracted for within 10 years after the creation of a zone, on any buildings or fixed plant and machinery which is an integral part of the building, benefit from **enterprise zone allowances**. These are similar to capital allowances, but the taxpayer can claim an initial allowance of 100% of his capital expenditure immediately.

If you dispose of an industrial building, or building within an enterprise zone, for which capital allowances have been claimed, and the disposal price is greater or less than the amount of expenditure for which allowances have been given, then a balancing charge or allowance will be made against income profits. The aim of the balancing charge or allowance is to ensure that total capital allowances given reflect the actual depreciation of the asset between acquisition and disposal. In practice, capital allowances for plant and machinery are pooled, so that on disposal of a particular item any balancing charge or allowance will be pooled with other capital allowances and may therefore not have an immediate effect on income profits.

To the extent to which the disposal proceeds exceed the original cost of the asset (plus other allowable expenditure), in addition to any balancing charges relating to capital allowances, there will be a capital gain chargeable to capital gains tax or corporation tax.

1.6.2 VAT

VAT is a turnover tax charged on most supplies of goods and services made in the UK in the course of business. It is normally borne by the ultimate consumer of the goods and services. The tax is levied at each stage of the production and distribution chain on the basis of the value added at that stage to the goods and services. The value added is the difference between the input cost and output price of the product. Each person in the chain charges, and pays to Customs and Excise, VAT (output tax) on the value of the taxable supply he makes, and recovers, from Customs, VAT (input tax) paid by him on supplies made to him which are attributable to his own taxable supplies.

Supplies of goods and services can be divided into:

- exempt supplies (e.g. insurance, financial services, education, healthcare) – output tax is not chargeable and input tax cannot usually be recovered;
- zero-rated supplies (e.g. food, books, newspapers, transport) – output tax is chargeable at a zero-rate (consequently no VAT is charged), but input tax can be recovered;

- standard-rated supplies (all other goods and services except domestic fuel which is currently charged at a special rate) – output tax is chargeable and input tax can generally be recovered; and
- supplies which fall outside the scope of VAT (e.g. supplies made to non-EU residents, supplies made within a VAT group and transfer of a business as a going concern).

Where input VAT is charged to you, for example if you purchase a new building or a landlord has opted to tax (see below), the amount of the VAT which you can recover will depend on the nature of the supplies made in the course of your business, and the resulting VAT recovery rate. If you only make standard-rated or zero-rated supplies, you will normally be able to recover all input VAT included in your expenditure; if some of your supplies are exempt, you will only be able to make partial recovery of VAT. The **Capital Goods Scheme** aims to take account of variations in entitlement to recover input tax incurred on the acquisition of capital items, including property. The VAT recovered will be adjusted to reflect the changing circumstances of the business over the life of the asset. Where the proportion of exempt supplies being made by the business grows this will be disadvantageous, but where a business is moving towards a larger number of taxable supplies it may be able to recover more input VAT. For example, a bank increasing the amount of general investment advice in comparison with financial services would increase its recovery rate.

(a) Exempt supplies

In general, and subject to certain exceptions (see below) the grant of any interest in or right over property and the surrender of such an interest or right is an exempt supply for VAT purposes. Accordingly, the person granting the interest (for example, selling a freehold interest in land, or granting or assigning a leasehold interest) will not charge VAT on making that supply, but neither will he be entitled to recover any input tax paid on supplies made to him, such as repairs maintenance and costs of purchase and sale. Consequently, any such input tax will be a cost which he must bear himself or recover by increasing the price or the rent charged.

(b) Standard-rated supplies

There are a number of important exceptions to the above basic rule. If any of these exceptions applies, then the supply in question will be a standard-rated supply, and the person making the supply will be required to charge VAT when he makes the supply and account for it to Customs and Excise. Equally, he will be entitled to recover any VAT paid by him on input supplies which are directly attributable to the standard rated supply he makes.

Freehold of a new building

The sale of the freehold interest in a new building is a standard-rated supply for VAT purposes (as is the grant of any right to acquire the freehold, such as an option to purchase in the future). The seller must account to Customs for the appropriate amount of VAT and if the seller wishes to recover the VAT from the purchaser, he must ensure that his contract for sale expressly provides that VAT will be charged in addition to the agreed purchase price.

Since the sale is a standard-rated supply, the seller will have the advantage that he will be able to recover all the VAT paid by him on supplies made to him in relation to the land and building concerned.

A building is new building if it was completed less than three years before the sale of the freehold, and a building is regarded as being completed when a certificate of practical completion is issued, or when it is first fully occupied, whichever happens first.

An important point to note is that only the disposal of the freehold of a new building is standard rated and if a new building is disposed of by the grant of a long lease, this will be an exempt, rather than a standard-rated supply, unless the person granting the long lease has exercised the option to tax (see below).

Option to tax

A person granting an interest in or right over land may elect to waive the exempt status of the property. This is commonly known as **VAT election** or **opting to tax**. This makes the grant of any interest in or right over the property by the person opting to tax, or any company in the same VAT group (see below), standard rated. This enables input VAT paid by the person opting to tax on expenditure relating to the property to be recovered and is particularly attractive where substantial expenditure on the property is expected, such as where the property is to be refurbished.

Once the option to tax is made, all payments made by a tenant under a lease, including rent service charge and insurance, will normally be subject to VAT. The landlord will usually be entitled to recover this VAT cost from the tenant under the terms of the lease. If the tenant makes only standard-rated and zero-rated supplies as part of his business, the VAT paid by him on rent, service charge and other charges will be recoverable by him if it is attributable to such supplies. However, if the tenant makes only or mainly exempt supplies, the VAT on rent and other charges will not be recoverable and will be a cost to him. Consequently, whether a landlord opts to tax the property may depend on the type of supplies made by his prospective tenants and landlords of properties likely to be let to tenants with low recovery rates generally choose not to elect, for example for properties in banking locations.

If a tenant can recover all his input tax, he may benefit from the option to tax. The **service charge** is charged to tenants to reimburse the landlord for his expenditure on running the property. If the landlord has not opted to tax the property, he will not be able to recover any VAT which he pays to suppliers and the service charge cost passed on to tenants will include this cost. This element of the service charge cannot be recovered, even by tenants making taxable supplies, as it is included within the main service charge. On the other hand, where the property is VAT elected, the landlord will be able to recover his input VAT paid to suppliers and can charge the tenant the net cost of the services plus a separate VAT charge, which can be recovered by tenants making taxable supplies.

Where the landlord operates a **sinking fund** and has opted to tax, he will invariably require the tenant to contribute an amount in respect of VAT, even though if the tenant owns the fund VAT is not due until the landlord draws on the fund. Until this time the VAT element will not be recoverable by the tenant. Alternatively the landlord may choose to charge the VAT at the later stage, in which case you may need to budget for these payments.

In summary, opting to tax can benefit tenants who can recover VAT but usually adds to the occupancy cost of tenants who cannot recover all or part of their input VAT.

If you grant a **sublease** to another occupier, you should consider whether it would be beneficial to opt to tax the property. You will normally wish to do so if your landlord has elected as in this case you will have to pay VAT on the rent paid to him. You will need to consider whether the subtenant can recover VAT as otherwise the VAT will represent an additional cost to him.

Assignments

The assignment of a lease is treated in the same way as the sale of a freehold. If however a reverse premium is paid to the new tenant to accept the assignment of a lease, the transaction is treated as a standard-rated supply by the new tenant and VAT is payable on the premium by the person assigning the lease.

Incentives

Where an incentive, such as a **capital payment**, is given by a landlord to a tenant, to enter into a lease, the tenant will be treated as making a standard-rated supply of services to the landlord. The same applies where the landlord pays the tenant for carrying out building or refurbishment works. Conversely, a **rent-free period** is usually ignored for VAT purposes unless it is expressed to be payment for the acceptance of the lease or is directly linked to anything else, such as works or other services undertaken by the tenant. Where VAT is chargeable on incentives, the VAT will be charged to the landlord and he can only recover his input tax if he has opted to tax the property.

(c) Special VAT provisions

There are special regulations that apply to the transfer of assets as part of a **transfer of a business as a going concern**. Subject to the performance of certain conditions, VAT is not charged on the transfer. The disposal of an interest in a tenanted building may constitute the transfer of a business or part business under these provisions.

In certain circumstances related companies can form a **VAT group**, supplies within the VAT group are not subject to VAT, so in some cases VAT can be avoided by grouping companies into a VAT group.

Supplies in relation to **protected buildings** are zero-rated and as such input tax incurred in respect of works to a listed or other protected building can be recovered.

1.6.3 Stamp duty

Stamp duty is a tax on certain documents which are executed in the UK or which relate to UK property or transactions. Because the duty is charged on the document itself, and not the transaction, stamp duty can usually be avoided where a transaction can be effected without the need for any written documentation, or where the documents are executed and held outside the UK. A written document is almost always required to transfer a title in land but in some unusual cases, such as surrender of a lease by operation of law, a document can be avoided (see Section 5.1.7).

Stamp duty is payable by the person to whom the interest in land is transferred or granted. It is payable on the following property transactions.

- Sale of a freehold. Stamp duty is charged at a rate of 1% of the price for a conveyance or transfer on sale of property, unless the price is £60 000 or less (and the transfer document can and does contain a certificate of value stating that the transaction does not form part of a larger transaction or series of transactions for a total price in excess of £60 000) in which case no stamp duty is payable.
- Assignment of a lease. Stamp duty is payable at 1% on the premium paid for the assignment of a lease in the same way as for the sale of a freehold.
- Grant of a lease. Stamp duty is charged upon both any premium payable on the grant of a commercial lease and on any rent payable under the lease. For stepped rents and leases including rent-free periods an average rent is calculated.

  Stamp duty is charged at a rate of 1% of the premium, unless the premium is £60 000 or less and the average annual rent payable under the lease is £600 or less and the transfer document contains an appropriate certificate of value in which case no stamp duty is payable on the premium.

  Duty is payable on the average annual rent payable under a lease at an increasing rate between 0% and 24%, depending on the average annual rent and the length of term of the lease. For example, the duty on a lease for 8–35 years at a rent greater than £500 is 2% of the annual average rent. There are various methods of mitigating duty on long leases.

A licence is not subject to stamp duty.

- Agreement for lease. Stamp duty is payable on an agreement for lease as if it were a lease. Duty is also payable on the lease when it is granted, but credit is given for the duty previously paid on the agreement.

- Surrender of a lease. Where a lease is surrendered, and this is evidenced in writing, stamp duty is charged at a rate of 1% of the market value of any payment made by the landlord for the surrender of the lease. If there is no payment for the surrender, fixed rate stamp duty of 50p is charged. The need for a document can be avoided where a lease is surrendered by operation of law (Section 5.1.7). Although this stamp duty is payable by the landlord, he may try to pass this cost on to the tenant as a term of the surrender.

- Exchanges. Where the payment for the transfer or grant of an interest in property includes another property, then the amount of stamp duty payable depends on how the transaction is characterised. If the transaction is an exchange of one property for the other, then both transfer documents are subject to 1% stamp duty, calculated by reference to the value of one of the properties. The documents must be carefully drafted to avoid both documents being stamped by reference to the more expensive property.

  The double tax charge of an exchange can be avoided if the transaction is structured as the sale of the more expensive property, the price for which is met partly by cash and partly by the transfer of the other property. Then the document transferring the more expensive property is chargeable to 1% stamp duty on its purchase price (which will include the value of the other property), whereas the document transferring the other property is only subject to a fixed duty of 50p.

In all of the above transactions, if part of the price (including rent) for the transaction is uncertain at the date of the execution of the document giving effect to it, the Inland Revenue will charge stamp duty by reference to the maximum amount which might become payable and which can be calculated on the date of the execution of the document. Where the price is totally unascertainable then stamp duty will be charged by reference to the market value.

The amount upon which stamp duty is charged also includes any VAT on the transaction. Moreover, where a landlord has the option to charge VAT on a property but has chosen not to do so, unless there is a binding agreement between the landlord and the tenant that the option will not be exercised, stamp duty is payable on leases as if VAT were charged on the rent.

There are certain reliefs from duty available for transfers within groups of companies and on certain corporate reconstructions.

1.6.4 Business rates

Business rates are normally a significant proportion of occupancy cost and in some cases can be larger than any other single component. Managing business rates effectively is therefore central to occupancy cost management. This section outlines the framework for calculating and paying rates and will assist you in challenging and controlling rates costs.

(a) Rating liability

Non-Domestic Rates are an annual tax on commercial property. In general the occupier, or the person who has the legal right to occupy, pays the non-domestic rates. Where the property is occupied, the occupier is obliged to pay the rates but the landlord may contract to do so on his behalf. Where the property is vacant, the person entitled to possession of the property is liable for the rates.

Despite being a central government tax, rates are collected by local authorities, which are district councils, London boroughs, unitary authorities and metropolitan councils. The relevant bodies are called billing authorities and have duties to collect all legally due rates.

Rates can be paid annually, or monthly over a ten-month period. In the event of an overpayment of rates, interest is payable. They can be paid by standing order or upon demand. If they are paid late, payment by instalments may not be allowed and interest on overpaid rates may not be paid.

To be liable for rates there has to be a rating assessment on a **hereditament** shown in the **rating list** and the property has to be occupied, or capable of being occupied. A hereditament is a property which is capable of separate assessment for rating purposes. It can be a whole building, or part of a building, or land or rights over land. Commercial properties can be either non-domestic hereditaments or composite hereditaments. A composite hereditament is a property which is part domestic and part non-domestic.

In the case of a new building or a building undergoing structural repairs, liability to pay vacant rates is normally triggered by a **completion notice** which can be served once the work remaining to be done can reasonably be expected to be completed within three months. The completion notice specifies the completion day, which is the later of the date of the notice and the date the building can reasonably be expected to be completed. There is a right of appeal against a completion notice.

A hereditament must be capable of definition. The easiest way to consider this is that it must be possible to draw a line around the property. For example, a market stall which has no fixed location will not be a hereditament (although the market as a whole may be), but one in a fixed location will be.

There are four tests for judging rateable occupation:

- the occupation must be **beneficial**. This does not necessarily mean profitable; it can satisfy a need, as in a public library or a hospital;
- The occupation must be **exclusively for the purposes of the occupier**. There can be more than one rating hereditament in the same location but the use of each hereditament must be exclusive;

- The occupation needs, by its nature, to be **non-transient**. Occupation of an office building for six weeks would be rateable because the nature of an office occupation is permanent. Conversely, builders' huts, if on site for only a very short time, are transient and not rateable.
- There needs to be **actual occupation**, but if a building is vacant but is retained for future occupation, it may in certain circumstances be treated as occupied;

Where premises are not occupied, the person entitled to possession is liable for **vacant rates**. This liability is usually 50% of occupied rates following a three-month void rate period.

If part of a property is vacant it may be possible to have it separately assessed and therefore benefit from 50% or 100% rate relief. An alternative is to have the rateable value apportioned by the valuation officer between vacant and occupied parts. Vacant rates are then charged on the relevant rateable value apportionment. This is a voluntary arrangement at the behest of the billing authority. It is most commonly used in industrial areas where occupation of factories varies with the order book.

If a property is vacant but only receiving 50% relief, it may be possible to obtain a reduction to a nil value by 'constructive vandalism' or 'soft stripping'. This means rendering the property incapable of beneficial occupation by removing or destroying key elements such as stairs or services. The scope of the work should be agreed with the valuation officer and normally with the landlord if the property is leased.

There are a number of additional reliefs and exemptions, including the following.

- 100% relief is available for unoccupied premises in some cases including:
    - industrial, storage, mineral or electricity generating properties.
    - if occupation is prohibited by law.
    - listed buildings and ancient monuments.
    - properties with a rateable value less than £1,500 (with the exception of advertising hoardings).
- Charities receive 80% relief where the hereditament is wholly or mainly used for charitable purposes.
- Agricultural land and buildings and fish farms are exempt.
- Certain property or parts of property are exempt if provided for the disabled or for people suffering from illness.
- All properties in enterprise zones are exempt until the end of the enterprise zone scheme (ten years after creation of the scheme).

Annual liability is calculated by multiplying the **Rateable Value** by the National Non-Domestic Rate, which is normally referred to as the **Uniform**

Table 1.7 Uniform business rate multipliers

Year	England	Wales
1990/1	0.348	0.368
1991/2	0.386	0.408
1992/3	0.402	0.425
1993/4	0.416	0.440
1994/5	0.423	0.448
1995/6	0.432	0.390
1996/7	0.449	0.405

Business Rate (UBR) and is set by central government. The appropriate rates in England and Wales since 1990 are shown in Table 1.7.

The UBR increases by the rate of inflation each year except in a year of revaluation. In those years the UBR is calculated so that the total rates collected is constant in real terms. The result is that, between revaluations, business rates should only increase by the rate of inflation. At revaluation the total amount of business rates payable remains constant in real terms but is re-apportioned to reflect the relative changes in value between properties.

Rates, although collected locally, are a national tax. If a local council desires additional funds it has to find these from other sources. The City of London has the right to levy an additional rate, but has not done so.

The simple calculation of rate liability is complicated by **transitional arrangements**, often known as **'phasing'**. Every five years, valuation officers around the country create new rating lists with a new valuation date. Each time that this has happened (since the Scottish Revaluation in 1985) the government has decided to dampen the effect of the revaluation by introducing transitional arrangements. The last revaluation was effective from 1 April 1995.

Under the current transitional arrangements, rates are calculated by reference to the rates payable in the previous year. The calculation of rates payable, simplified, is as follows.

- For properties which are subject to increases in liability due to the 1995 revaluation, the lower of (a) the rateable value multiplied by the UBR and (b) the last year's rate liability plus inflation multiplied by the appropriate phasing multiplier.
- For properties which are subject to decreases in liability due to the 1995 revaluation, the higher of (a) the Rateable Value multiplied by the UBR and (b) the last year's rate liability plus inflation multiplied by the appropriate phasing multiplier.

The multipliers are shown in Table 1.8.

Table 1.8 Phasing multipliers

Year	Multipliers where increases apply			Multipliers where decreases apply	
	Large properties (%)	Small properties (%)	Small composite properties (%)	Large properties (%)	Small properties (%)
1995–6	110.00	107.50	105.00	95.00	90.00
1996–7	107.50	105.00	102.50	95.00	90.00
1997–8	110.00	107.50	105.00	85.00	80.00
1998–9	110.00	105.00	105.00	70.00	65.00
1999–2000	110.00	107.50	105.00	70.00	65.00

A small property is one with a Rateable Value of £1 000 or less in London, £10 000 or less elsewhere. All other properties are large properties.

(b) Assessing the rateable value

The rateable value is defined as:

> an amount equal to the rent at which it is estimated the hereditament might reasonably be expected to let from year to year if the tenant undertook to pay all usual tenant's rates and taxes and to bear the cost of the repairs and insurance and other expenses (if any) necessary to maintain the hereditament in a state to command that rent.

It is therefore the **rental value** for an annual tenancy on effective full repairing and insuring terms. In practice, this is normally similar to the rental value for fixed-term leases.

The rental value is established using rental evidence on the property itself or comparable properties. Other assessments can also be used as evidence of value, commonly referred to as the **'tone of the list'**. All rents have to be adjusted so that they reflect the statutory definition. It has to be assumed that the hereditament is vacant and to let. This does not mean that the occupier is excluded from the theoretical bidding for the property; he is deemed to be in the market and one of the hypothetical tenants who come fresh to the scene and to have no preconceived ideas about the property or its location.

For some types of properties, there is no readily available evidence of rental value. In these circumstances alternative methods of valuation are required to arrive at the rateable value. These include:

● profits test – this test applies to properties which have some form of locational monopoly, such as petrol stations, hotels, cinemas, etc. Accounts are analysed to determine the rent that can be supported by the business;

● contractor's test – this is similar to the Discounted Replacement Costs basis of valuation (Section 1.4.2). It is derived from the capital cost of constructing a replacement building and buying land, discounted to reflect obsolescence and environmental factors. Once a capital value is derived it is decapitalized to an annual basis using a statutory decapitalization rate. In very rare instances the capital cost of construction and value of the land can be replaced by its capital value.

● statutory formulae – these apply to statutory undertakers and some specialist uses such as British Rail, London Underground, British Gas, canals, railways, docks and harbours.

The **valuation date** (known technically as the antecedent valuation date) is the 1 April 1993 for the list in force from 1 April 1995 and for the previous list was 1 April 1988. Rental value is assessed as at the valuation date, but the physical and environmental state of the property is taken at the time the rating list comes into force (for the current list, 1 April 1995). Specifically, the following **mentioned matters** have to be taken as they exist at the date the list comes into force:

● matters affecting the **physical state or physical enjoyment** of the premises, such as additions, alterations and the condition of the premises;

- the mode and category of **occupation** of the premises, i.e. the use to which the property is put;
- matters affecting the **physical state of the locality** in which the premises are situated or which, though not affecting the physical state of the locality, are none the less physically manifest there.

  The state of the locality is a relatively simple concept, but the 'physical manifestation' is less so. An example of 'physical manifestation' might be Prestwick Airport which was by law the only international airport in Scotland. This law was repealed and Prestwick ceased to be the major airport. Despite no physical changes on the ground, the change of status was 'physically manifest' by fewer planes coming and going;

- the **use or occupation of other premises** situated in the locality of the premises. This can have a marked effect on value, for example, if a major office occupier moves into a vacant office development;
- the quantity of **minerals** and other substances in or extracted from the premises;
- the quantity of **refuse** or waste material which is brought onto and permanently deposited on the premises.

Additionally, if there is a change in any of these mentioned matters after the rating list comes into force, the change will be taken into account in assessing the rateable value. For example, an extension added on to a building after the valuation date will be taken into account in ascertaining the rateable value as if the extension had existed at the valuation date.

(c) Altering the rating list

The most common way of seeking to reduce rate liability is to make a **proposal** to alter the rating list, which, if not accepted, operates as an **appeal** against the rateable value assessment.

The list is maintained by the **valuation officer** (VO) for the area. The valuation officer can alter the rating list at any time. Other interested parties, such as the occupier, the owner and in certain circumstances the rating authority now have the right to make a proposal at any time during the currency of the list and for a period afterwards (usually one year).

Grounds for making a proposal include:

- an inaccurate valuation;
- an incorrect effective date;
- a change in one of the 'mentioned matters' discussed above;
- the hereditament should not be in the list at all;
- an inaccurate description;
- other specified matters relating to the accuracy and fairness of the rating list.

Following a revaluation, it is important that you consider your rating assessment carefully to determine whether an appeal is appropriate and it is usually advisable to take professional advice at this stage. Other important times to consider appeals are at a change in ownership or occupation, when there are construction works in the premises nearby, or if the premises fall into disrepair.

In some cases, phasing arrangements will mean that it is not worth making an appeal, as the reduction in rates payable will be limited or may take several years to come through. However, you should remember that the assessment will probably affect the base level from which phasing is calculated after the next revaluation, so even if you do not see the benefit of a reduction immediately there may be a significant long-term benefit.

The grounds for a reduction in the rateable value can include:

- over-valuation of the property or a class of property;
- demolition of part of the premises;
- disrepair;
- adjacent development (for example, for retail property, the opening of a new shopping centre);
- changes in the law affecting the way the property is used or occupied;
- substantial changes in the occupation of other premises in the locality;
- change of use of the property.

Values can sometimes be increased by the valuation officer. This can happen, for example, when the valuation officer corrects an undervaluation of the property, or improvements are carried out to the property or in the locality.

Once an appeal is launched the valuation officer has four weeks to judge whether it is valid and serve an **invalidity notice**, which can be disputed through the valuation tribunal (see below). Alternatively, the valuation officer may agree with the proposed alteration and alter the list accordingly. More commonly, the valuation officer will seek to negotiate and reach agreement. If the case is not settled within six months, it is referred to the **Valuation Tribunal**. In practice, the valuation officer may not start negotiating until the case has been listed for hearing.

If agreement cannot be reached the appeal will have to be heard by the valuation tribunal. The valuation tribunal will make its decision on the basis of the evidence presented to it. If any party who attended the hearing is aggrieved by the decision an appeal may be made to the **Lands Tribunal**. Such an appeal is a serious matter and can be expensive; it is normally only appropriate if the case is clear-cut and no compromise is possible, or a matter of principle is at stake. The lands tribunal is the last court of valuation. An appeal from there is on points of law only to the Court of Appeal and thereafter to the House of Lords.

(d) Instructing advisers

The rating system is complex and if you are to protect yourself effectively and minimize your liability professional advice should be taken at all stages, particularly when you are considering an appeal.

In many cases a rating surveyor can advise, but in complex cases legal advice should also be sought, ideally from a lawyer who has specialist expertise in rating law. The rating surveyor should have an understanding of the legal framework, valuation expertise, market knowledge and negotiating skills. He should be qualified, and be regulated by a professional body: either the Royal Institution of Chartered Surveyors (RICS), the Incorporated Society of Valuers and Auctioneers (ISVA) or the Institute of Revenues, Rating and Valuation (IRRV).

Unqualified rating advisers have proliferated recently and occupiers should be wary of advisers approaching them aggressively and promising reductions in rates bills. If possible rely on a recommendation. The professional bodies provide a referral service. The Royal Institution of Chartered Surveyors also provides a rating helpline service which offers half an hour's free advice.

Fees are normally on the following basis (or a combination of them):

- a fee fixed as a percentage of the rateable value;
- a fee based on any reduction in the rateable value or rates payable achieved;
- a fixed fee based on the anticipated work;
- a time-based fee.

A fee basis including a fixed element and a contingency element is common. The occupier should consider carefully whether this type of fee structure is advantageous as in many cases, especially for properties with high rateable values, a fixed fee or time based fee is more cost effective.

FURTHER READING

ANAA (1991) *The Accounting Classification of Real Estate Occupancy Costs.*

Apgar, M. (1991) Reducing corporate occupancy costs: A real estate opportunity, *Urban Land,* 16 September, Land Institute, Washington DC.

Apgar, M. (1993) Occupancy costs, *Harvard Business Review*, May/June.

Bernard Williams Associates (1995) *Facilities Economics*, Building Economics Bureau Limited, Bromley, Kent.

Bogan, C. E. and English, M. J. (1994) *Benchmarking for Best Practice – Winning Through Innovative Adaptation*, McGraw-Hill.

Britton, W., Davies, K. and Johnson, T. (1980) *Modern Methods of Valuation,* The Estates Gazette, London.

Davidson, A. W. (1989) *Parry's valuation and investment tables*, The Estates Gazette, London.

Debenham Tewson Research (1992) *The Role of Property – Managing Cost and Releasing Value*, London.

Enever, N. and Isaac, D. (1994) *The Valuation of Property Investments*, The Estates Gazette, London.

Levy, H. and Sarnat, M. (1990) *Capital Investment and Financial Decisions*, Prentice Hall International (UK), Hemel Hempstead.

Liebfried, K. H. J. and McNair, C. J. (1994) *Benchmarking: A Tool for Continuous Development*, Harper Collins.

Lumby, S. (1991) *Investment appraisal and Financing Decisions*, Chapman & Hall, London.

Maas, R. W. (annual) *Tolley's Property Taxes*, Tolley Publishing Company Limited.

Mott, G. (1993) *Investment Appraisal*, Pitman Publishing, London.

Nourse, H. O. (1989) *Managerial Real Estate*, Prentice-Hall, New Jersey

RICS (1995) *RICS Appraisal and Valuation Manual*, RICS Business Services Ltd, London ('*The Red Book*').

RICS (1996) *Guide to Securing the Services of a Chartered Surveyor: Valuation, Property and Estates Management*, RICS Books, London.

Scarrett, D. (1995) *Valuation: The Five Methods*, E. & F. N. Spon, London.

Silverman, R. A. (1987) *Corporate Real Estate Handbook*, McGraw-Hill Book Company, New York.

Touche Ross (annual) *The Financial Reporting and Accounting Manual*, Butterworths, London.

2 | Acquisition

KEY POINTS

- The structured approach to acquisitions
- Assembling a strong professional team
- Negotiating strategy and tactics

One of the most significant single event a business faces is relocation into new premises. Not only does the process itself dominate during the acquisition and relocation period, but the premises chosen and transaction structure negotiated will have a significant impact on business efficiency and profitability in both the short and long term.

Yet, often the business manager has little experience in the area and is ill-equipped to manage the process effectively. The acquisition process can be complex and requires skills and knowledge in areas such as business planning, finance, law, accounting and negotiation. This chapter aims to provide an appreciation of how to manage the process to achieve the best possible transaction to meet the ongoing needs of the business. The information and advice contained in the chapter provide the occupier with a foundation for identifying requirements, managing advisers, and negotiating the transaction with an informed and controlled approach. You should also obtain a copy of *Commercial Property Leases: Code of Practice*, listed under Further Reading. This is an important document containing both useful information and guidance on recommended practice. You should familiarize yourself with the contents before embarking on an acquisition project.

The business occupier is not normally a specialist property professional. Even where property is managed by a specialist in-house team, the focus of resources and expertise is often on the company's core business activities. In contrast, the property owners with whom the occupier will be negotiating for an acquisition are usually operating in their core business area. They are normally property professionals with significant experience in negotiating transactions. As a result of this imbalance occupiers often fail to achieve the most favourable transaction structure and the one that most effectively meets their business needs. The landlords and their advisers take control of the process and the occupier ends up with a transaction imposed and not chosen.

It is rarely possible to structure a relocation which represents a perfect fit with business needs. What can be achieved in a particular case depends on the properties available and relative negotiating strength of the parties. The aim of the acquisition process is to achieve a transaction which represents the best possible fit available at the time in the market. In order to accomplish this the occupier must take control of the process and follow a structured approach. This approach can be more costly in the early stages, but it saves time and money overall.

The essential characteristics which differentiate the structured approach are:

- a carefully planned and controlled approach at all stages;
- a comprehensive and coordinated professional team fully involved at all stages.

Figure 2.1 shows all the stages of the structured acquisition process. This is a simplified programme and in practice a number of the stages will overlap. In total the process is likely to take a minimum of six months and can take as much as three years in very complex cases. In some cases there will be overriding business reasons making it essential to cut short the programme, and in these cases it may be justifiable to make a sacrifice in terms of the transaction negotiated.

The approach enables the occupier to govern the process and make choices consciously and in accordance with business objectives. Each stage in the process is essential and bypassing or curtailing any stage, which often happens as a result of time or cost constraints, can have detrimental implications in the long or short run. Too often occupiers launch into the property search and negotiation before they have established the foundations by analysing objectives and requirements properly and assembling the professional team.

It is only possible to follow the structured approach if enough time and resources are devoted to the project. All too often occupiers leave a relocation until it becomes urgent and there is insufficient time or resources to maximize negotiating strength and the effectiveness of decision making. Time constraints mean the process has to be compressed at each stage. Lack of resources, such as restrictions on professional fees, can prevent effective implementation of each part of the process. This can translate into real costs which may significantly outweigh any savings.

The timescales shown in Figure 2.1 are approximate estimates and reflect a range of types and sizes of transaction. The time involved in a particular case depends on the nature of the requirement and the transaction, but it is normally unwise to leave less than a year for any acquisition project even for a small transaction. You may be able to complete some of the stages in a shorter time, and in some cases you may have no choice but to short cut the process where accommodation is required urgently, but ideally you should allow at least a year.

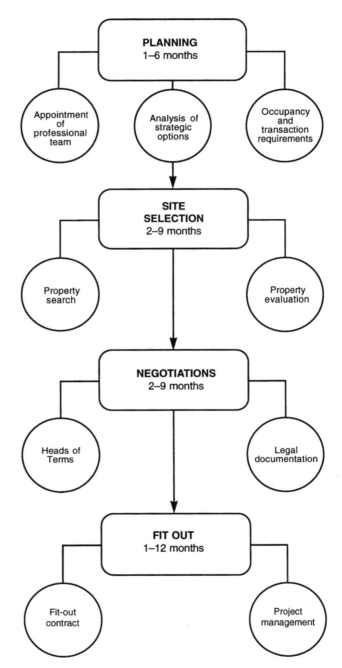

Figure 2.1 The structured acquisition process.

2.1 PLANNING THE ACQUISITION

This stage governs the effectiveness of the whole of the rest of the process but is regularly neglected. By clearly defining objectives and requirements, translating these into an effective strategy and assembling a professional team capable of implementing the strategy, the foundations are laid for a successful acquisition.

The time involved is often a significant portion of the total acquisition period. When an occupier is faced with pressure to move quickly, this part of the process often suffers. It can sometimes seem to be time wasting and the manager wishing to be proactive and efficient can be nervous of spending time apparently achieving little. But the time is not wasted and the work undertaken at this stage, though it may not be highly visible, will save abortive work later and can be crucial to the success of the project. Ultimately, investing the time at this stage can reap significant time and cost savings later. Not only will the property and transaction meet business needs effectively, but this will be achieved without costly and time consuming changes of direction.

2.1.1 Appointment of professional team

The quality of the professional team and the client's ability to manage the team effectively are very significant for the success of the project. Almost all acquisitions will involve some professional advice even if this is confined to the lawyer. It is usually advisable to instruct an acquisition agent to find the premises and negotiate the transaction. The agent's knowledge of the market and of the terms negotiable is essential to achieving the most appropriate property on the most favourable terms. In most cases, including small transactions, cutting back the professional team to save money is a false economy. What is saved in professional fees can be lost several times over in the transaction.

The only way to be certain of acquiring the best property on the most favourable terms and to be sure that the business is protected from unnecessary long-term risks, or at least to know that you are properly alerted to them, is to ensure that you are supported by a competent team of advisers. Obviously the size of the team must be tailored to the size of the project and a full complement of advisers is not always appropriate, but careful judgement is required in determining the size of the team.

It is tempting to use in-house staff when they are available and in some cases this is fine. For large organizations with property and legal departments, in-house resources are often ideally suited: the in-house professional has the advantage of a full understanding of the company's objectives. On the other hand, it is important to ensure that the in-house team has the necessary specialist expertise; if the right specialist skills are not available the success of the project could be prejudiced.

Be careful not to wear too many hats.

Figure 2.2 shows the structure of the professional team which would be relevant for most acquisitions of commercial premises.

The team depends on the size and type of project and the team structure shown in Figure 2.2 will not apply in all cases. For example, for construction of a new building or a major refurbishment a larger technical team would be necessary. Conversely, where premises are fitted out by the landlord construction advice can be streamlined. Sometimes the strategic analysis, process management and other functions can be undertaken in-house, but the in-house resources and expertise should be scrutinized carefully to ensure that these crucial elements of the process receive the emphasis and skills they require. In many cases one adviser will perform a number of roles. For a small project the agent will act as property adviser and process manager as well as conducting the search and negotiations.

Instructing advisers is discussed generally in Section 1.3 so please refer to that section for an appreciation of the general principles which apply, including issues such as choice of advisers, fee structures and preparation of instructions.

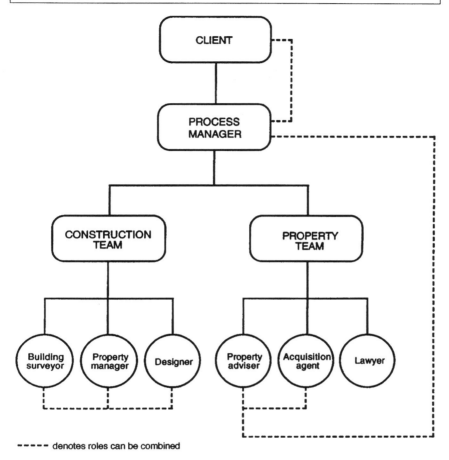

----- denotes roles can be combined

Figure 2.2 The acquisition team.

(a) The property team

● The process manager. It is important that one of the team members has the role of controlling the process and orchestrating the team. This is the key to ensuring the team works together effectively to achieve the objectives. The process manager can be an in-house manager or an adviser but in either case it is vital to ensure that he has the time and expertise to do justice to the job. The job is time consuming and requires an understanding of the roles of each of the advisers and of the process.

The process manager is responsible for coordinating the team to ensure that each team member undertakes his role effectively. At the site selection stage, for example the process manager will coordinate the brokerage function and property advice and introduce the legal and construction teams as and when necessary. Similarly, at the negotiation stage the process manager coordinates the property and legal teams with support from the construction team to ensure that the most effective transaction structure is negotiated.

If you choose to instruct a process manager and not to manage the process in-house you may give the function to the property adviser or acquisition agent or you can instruct a separate consultant, possibly a professional that manages other projects for your business, such as a management consultant or accountant.

The process management function will normally require a time-based or fixed fee based on the anticipated time involved. Where the role is combined with negotiating or advising on the transaction the fee will be included with the main fee, but you should ensure that the fee is adequate to cover the work involved so that this central function is not neglected.

● The property adviser. The property adviser can be the process manager and acquisition agent but it is not necessary to combine the roles. It is usual to combine the functions but the result can be that the quality of the property advice and of the process management suffer. This is caused by two main factors. Firstly, there is a danger that the agent paid on a commission fee will be overly concerned to complete the transaction and may compromise objectivity. Secondly, the broker rarely quotes a fee to include significant advice and process management work and cannot therefore devote the necessary time to this function.

The property advice function comprises strategic analysis in the planning stage and advice on property issues throughout the process. In the planning stage the property adviser helps in the development and analysis of strategic options and in translating these into occupancy and transaction requirements. In the site selection stage, the property adviser assists in the evaluation process commenting on property issues as appropriate, such as advising on location, specification, cost, management and legal issues. During negotiations the property adviser supports the agent and lawyer and during fit out supports the project manager and advises on landlord and tenant issues.

The property adviser can be a surveyor or consultant with appropriate skills and expertise. The fee will normally be a time-based or fixed fee.

● The acquisition agent. The agent should have experience in negotiating acquisitions, preferably of the size and type concerned. Knowledge of the local market is advisable but not always essential. Finding an agent with good professional and negotiating skills, an understanding of the client's objectives and ability to find workable solutions, and the ability to work effectively as part of a team are often more important than local knowledge.

The agent's role is to undertake the property search and develop an initial inspection shortlist. He accompanies the client on inspections and assists in development of the shortlist for negotiation. He conducts the Heads of Terms negotiations and supports the lawyer in the legal negotiations.

The fee basis can be ad valorem, time-based or a fixed fee. Ad valorem fees have the advantage that they relate to the size of the project but have the disadvantage that the agent may be motivated to recommend agreement on unfavourable terms. In general, the fee should relate to the time expected to be involved. A fixed fee obviously has the advantage of certainty but if the work needed turns out to be greater than expected you take the risk that the agent will cut corners.

- The lawyer. The lawyer should be involved in all stages of the process to advise on legal issues. At the site selection stage the lawyer works with the property adviser and agent to report on the landlord's covenant, building defects and management issues. At the negotiations stage the lawyer should work closely with the agent and property adviser in agreeing Heads of Terms.

Traditionally, the lawyer's role has been confined to legal documentation but significant value can be added by involving the lawyer from the early stages in the property evaluation and particularly in negotiating Heads of Terms. Many clients concerned to manage professional fees see the lawyer's involvement before legal drafting as unnecessary. Additionally, the attendance at meetings in the early stages of negotiations can be interpreted as adversarial.

However, while the lawyer plays a secondary role at the Heads of Terms stage, he should work closely with the agent and other professionals to ensure that all legal issues are properly addressed. At the very least the lawyer should be involved in drafting counter proposals and advising on transaction structure and terms, and ideally you should not settle any terms without the lawyer's input. Additionally it is often advisable to be joined by the lawyer in Heads of Terms meetings.

The potential cost of not fully involving the lawyer can significantly outweigh the savings in legal fees. Without the lawyer's involvement in the early stages the terms agreed often need to be significantly revised at the legal stage when the tenant's negotiating position is at its weakest. All too often negotiating successes achieved at the Heads of Terms stage are subsequently lost at the drafting stage because of the renegotiation required as a result of a failure to be precise.

Legal fees are usually on a time-based, fixed, or occasionally a capped basis.

(b) The construction team

- The building surveyor. The building surveyor undertakes the building survey, sometimes supported by a structural engineer and a mechanical and electrical engineer. He advises on building defects, repair and service charge issues. He works closely with the agent, property adviser and lawyer in incorporating appropriate terms into the transaction and in preparing and reviewing technical documentation such as schedules of condition and base building specifications.

The building surveyor is normally a member of the Building Surveying or General Practice division of the RICS. The agent, property adviser or lawyer may be able to provide a recommendation. The fee basis is normally time-based or fixed fee.

It is important that the building surveyor's remit is clearly defined at the start of the project to avoid problems later on. Be sure that there is no misunderstanding over the brief, as disputes often arise where instructions have not been clearly established at the start. The surveyor will specify limitations on the survey undertaken, and you will need to consider these carefully to ensure that the validity of the survey will not be compromised. You should give him as much information as possible about the project and the proposed terms of the transaction so that he is aware of the important issues and can play a full role in the team.

- The project manager. In the planning stage the project manager has a role in providing cost and other information needed in the analysis. He assists in preparation of the occupancy requirements in the planning stage and in refining these in the subsequent stages. He manages the designer and space planner to develop space requirements. During the site selection and negotiation stages he undertakes a technical evaluation to appraise potential properties with respect to design, use and fit out and assists the agent and lawyer in incorporating appropriate terms into the transaction, including landlord's contributions to the fit out. He deals with the technical aspect of the documentation, particularly the licence to alter (Section 3.4.7). The project manager's primary role is at the fit out stage when he coordinates the process, advises on contracts and manages the designer and contractor (Section 2.4.1 below)

At the construction stage the project manager is an expert representative of and adviser to the client, monitoring progress, communicating decisions to the project team and keeping the client informed. The project manager will give advice on matters relating to the construction team, including contractors and subcontractors. It is important that the project manager is used to manage the fit-out team. This ensures that decision making is driven by business planning and cost decisions, and not the reverse. A construction team driven by the designer runs the risk that space planning and design is not governed by financial considerations and that requirements drive the process and determine cost.

The project manager should be a qualified professional such as a quantity surveyor, building surveyor, architect or specialist project manager. It is advisable to use a firm that has quantity surveying expertise to ensure that cost is managed effectively.

The fee basis can be a fixed fee calculated as a proportion of the construction cost or, less commonly, a percentage of cost or time-based fee. The RICS has a recommended scale fee for this work.

- The designer. In traditional construction procurement, the designer can be an architect, designer or surveyor. In some cases it is most cost effective to use the project manager for this work if they have design expertise. In the design and build contract the design is undertaken by the contractor (Section 2.4.2 below). The client should take the advice of the project manager on the appropriate professional to do this work.

  The designer assists the project manager and client in developing fit-out requirements based on the occupancy requirements prepared in the planning and subsequent stages. He then translates these into a design which will be further modified and developed during the course of the project.

  The fee for the designer is usually based on a percentage of the contract cost, although it can be a lump sum or based on time.

The construction team described here is the structure commonly used for a fit-out project where a traditional contracting route is adopted. Other procurement routes will involve variations on the team (Section 2.4.2 below). The structure is not rigid and in many cases will need to be adapted to accommodate the particular circumstances and the occupier's needs. In particular, for very small or very large projects the structure described may have to be altered. For example, a large contract may require a building surveyor, project manager, architect, quantity surveyor, mechanical and electrical engineer, and structural engineer, each of which will have a different role. Conversely, for a Design and Build contract the contractor takes on most of the responsibility for design and management and there will not normally be a need for a separate designer, or project manager.

It is essential that all team members have clearly defined roles and that they understand their responsibilities and terms of engagement. Often a letter of instruction is adequate, but for large or complex projects formal appointment documents are necessary, for example in cases where tightly defined obligations and responsibilities for the construction team are required. For large construction projects there are standard form contracts published by relevant professional bodies for the appointment of the professional team, including the architect, quantity surveyor, and engineers where these are needed for the project. As these standard forms are prepared by professional bodies they do tend, as a general rule, to favour the professionals. One of the main pitfalls of which to beware in such contracts is clauses which restrict or exclude liability.

2.1.2 Analysis of strategic options

The first step in the planning process is to define the strategic objectives which the business wishes to fulfil through the acquisition. These objectives will form the basis for developing detailed occupancy and transaction requirements which will be used in the site selection and negotiation stages.

Please refer to Section 1.1 where property strategy is discussed more fully. Examples of the types of objective which might be established in a particular case are:

- consolidate all staff in one location to maximize business efficiency;
- reduce occupancy cost to a particular proportion of turnover;
- achieve leasing flexibility, particularly options to expand, contract and break.

Figure 2.3 shows some of the types of issues which usually need to be addressed in determining strategic property objectives.

Strategic objectives can conflict with each other and the challenge is to manage an effective trade-off between objectives. For example, during the acquisition process itself time and cost constraints can mitigate against acquiring the property which most effectively achieves other business objectives, particularly long-term occupancy cost control and flexibility. Similarly, time constraints hamper the effort to meet objectives by preventing proper planning and site selection and weakening the negotiating position.

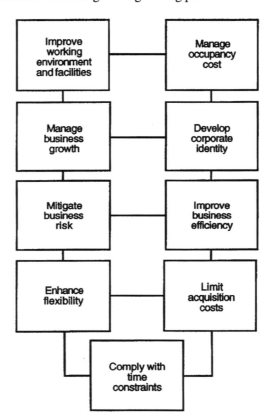

Figure 2.3 Acquisition – strategic objectives.

Once strategic objectives are defined the options which are thought to fulfil them can be identified. The options then need to be analysed qualitatively and quantitatively to establish a preferred strategy. The options may relate to alternative business strategies, such as consolidation versus diversification, to alternative locations, such as a central versus edge-of-town location, or alternative transaction structures, such as lease versus purchase. The analysis undertaken should follow the approach outlined in Section 1.2 and include consideration of the economic, accounting and business implications of each option.

The process of debating and analysing options is central to the overall acquisition process. It provides a strong foundation for moving ahead into the next stages. It is only through developing clear objectives and strategies that a property and transaction that meet the objectives can be identified. The process ensures that requirements are driven by the business plan and not vice versa.

The process prompts debate within the organization and the professional team which, having been resolved, enables the occupier to move ahead confidently with a coordinated approach. By rigorously analysing options, the business manager determines the appropriate strategy based not on instinct, 'gut feeling' or prejudice but on a careful analysis of the costs and benefits of each option. Not only does this give you confidence in your strategy, but it also provides a basis for convincing others in the organization and the professional team. This in turn helps to achieve the consensus essential if all involved are to be committed to the agreed strategy.

To demonstrate how the process works, Example 2.1 shows a fictional example of a choice between two locations. The net present values are calculated using a discounted cashflow technique as shown in Section 1.2.

Example 2.1 Locational options

In this fictitious example, a company has to decide between two broad locational options. This could be a choice between an edge of town location compared with the city centre, West End of London versus the City or any other locational decision.

Quantitative analysis

Net Present Value of 10-year lease of 5000 sq ft in Location A
  £3.1 million
Net Present Value of 10-year lease of 5000 sq ft in Location B
  £2.68 million:

Assumptions:
Discount rate: 9% (the company's average cost of capital.):
Location A: Rent £25 per sq ft, one year rent free, service charge £6 per sq ft, rates £19 per sq ft, relocation costs £50 000, fit-out costs £100 per sq ft, professional fees 10%, reinstatement costs £40 per sq ft, VAT 17.5%.

Location B: Rent £20, three years rent free, service charge £6 per sq ft, rates £19 per sq ft, relocation costs £50 000, fit-out costs £100 per sq ft, professional fees 10%, reinstatement costs £40 per sq ft, VAT 17.5%.

Qualitative analysis:

Location A has better transport links, communications and facilities.
Location A is better located for clients and staff.
Location A is better suited to the corporate image the company wishes to project.
Location A has a greater risk of rental growth.
Location B is better located for suppliers.
The availability of appropriate space is greater in Location B.

The company must weigh up the qualitative and quantitative issues in arriving at a decision on the preferred location.

Having evaluated strategic options and decided on the preferred strategy, the company now needs to establish the detailed requirements for the acquisition. These are formulated in the third part of the planning stage.

2.1.3 Development of occupancy and transaction requirements

The detailed requirements relate to the building and its management (**occupancy requirements**) and the structure and terms of the transaction (**transaction requirements**). By focusing on the detailed requirements in the early stages the occupier can retain control of the process through all its stages.

The occupancy and transaction requirements established in the planning stage can subsequently be refined and adapted during the later stages of the process. You will have to refine priorities in response to the attributes of the shortlisted properties and the terms the landlord is prepared to offer. However, ranking the requirements at the planning stage assists in decision making in later stages. Without this clear idea of priorities the occupier ends up making decisions later in the process in a hurry without proper consideration.

Occupancy requirements fall into three categories, locational, building and landlord issues. The types of issue to be addressed are shown in Table 2.1

By considering each of these issues a list of requirements ranked by level of importance can be developed. Example 2.2 is a fictitious example of such a list which demonstrates the types of requirement that might be established and the ranking system.

Table 2.1 Occupancy requirements issues

Locational issues	Building issues	Landlord issues
★ proximity and accessibility to existing staff and labour pool including skill and age profile of labour pool	★ existing building or new building	★ ownership structure
★ proximity and accessibility to existing and prospective clients and customers	★ size and style of building, appearance, image, visibility and position	★ landlord's financial status (current and projected financial strength)
★ proximity and accessibility to suppliers and competitors	★ building size and layout (capacity to accommodate existing and future space usage requirements)	★ landlord's management style (reputation, approach to landlord/tenant relationship, conciliatory/aggressive style)
★ access to public transport (rail, underground, air)	★ letting structure (single/multi-let) and other tenants	★ building management regime (staffing level, quality and approach, outsourcing to managing agents)
★ access to road network	★ condition/age	
★ facilities and amenities (leisure, entertainment and conference facilities, shops, restaurants, pubs, parks, sports facilities etc.)	★ facilities (parking, security, reception, dining and meeting rooms, storage)	
★ availability of local housing, schools and community facilities	★ natural light and outlook	
★ planning regime (approach of local planning authority to changes of use and to development proposals)	★ hours of access and access routes (entrances, stairwells, passenger/goods lifts)	
★ environmental considerations such as contamination problems and statutory obligations	★ marketability for disposal	
★ tax issues especially rates, enterprise zone allowances, VAT	★ base building specification and services (including e.g. power and communications facilities, heating and cooling, floor loadings, plant capacities, lighting, toilet facilities, structural grid and divisibility)	
★ total occupancy cost differentials (rent, rates, service charge and other operating expenses including facilities management, utilities and fit out)	★ state of repair and defects	
★ staff cost differentials (wages, housing, travel and relocation costs)	★ type of finish (shell and core/developer's finish/fitted out)	
★ other business cost differentials including transport, communications, supplies and other operating costs	★ planning constraints on alterations and use	
	★ compliance with building and health and safety regulations	
	★ energy efficiency	
	★ special technical requirements (standby generator, air conditioning enhancements, security systems, telecommunications)	

Example 2.2   Occupancy requirements – City of London office acquisition for a financial services company

Ranking options: Essential/Important/Preferred

Requirement	Priority ranking
Locational issues:	
Town centre	Essential
Within 3 minutes walk of Underground station	Preferred
Within 5 minutes walk of public car park	Preferred
Building issues:	
Style/condition:	
High profile existing building	Preferred
Multi-let to 'blue chip' tenants	Important
Top specification with impressive facade, entrance and common areas	Important
Good natural light and outlook	Preferred
Newly constructed or refurbished within 2 years	Preferred
No significant building defects exposure	Important
Accommodation:	
8000 – 10 000 square feet	Essential
Adjacent expansion space 2000 square feet	Preferred
500 square feet separate self-contained secure storage	Important
Facilities:	
On-site car spaces minimum 2	Important
24-hour access	Essential
24-hour manned reception	Important
Dining room for common use	Preferred
Meeting rooms for common use	Preferred
Exclusive use of toilets on floor	Preferred
Base building specification:	
Air conditioning	Essential
Raised floors minimum 100 mm void	Essential
Full access suspended ceilings	Essential
Passenger lifts minimum 2 with minimum 8 person capacity	Preferred
Landlord issues:	
Minimum landlord credit risk	Essential
Cooperative management approach	Essential
Efficient on-site management regime	Essential

Space requirements need to be estimated at this stage as part of the occupancy requirements. These can be estimated broadly using average space requirements per employee, but it is normally advisable for the project manager and designer to establish more accurate requirements based on the business plan and requirements of individual departments. Normally block plans will be prepared initially which map out the broad area to be used by each department.

All the occupancy requirements need to be linked to cost decisions to ensure that the amount, quality and location of space acquired is based on strategic requirements (Section 1.1 discusses affordability and the performance/cost trade-off). It is important that demand is managed so that the optimal quality and quantity of space, taking into account projected costs and revenues, is acquired to maximize profitability.

Transaction requirements are developed in a similar way to occupancy requirements. Please refer to Chapter 3 for a full discussion on transaction structure and terms and for an explanation of the terms used in Example 2.3. From that section the occupier should be able to establish the core transactional issues and hence a prioritized list of transaction requirements. Example 2.3 is a fictitious example of a transaction requirements list at the planning stage. Again, the list will be refined over the acquisition process, particularly in the site selection stage and the early part of the negotiation stage.

Example 2.3   Transaction requirements – City of London office acquisition for a financial services company

Ranking options: Essential/Important/Preferred

Requirement	Priority ranking
Term:	
Maximum term 25 years	Essential
Minimum term 3 years	Essential
Break option at every fifth year	Important
Contraction and expansion rights at every third year	Preferred
Occupancy cost:	
Total rent, rates and service charge maximum £50 per square foot	Essential
Net effective rent maximum £20 per square foot	Preferred
Rent fixed for 10 years without review	Preferred
Upwards/downwards rent reviews 5 yearly	Important
Independent measurement to verify floor areas	Important
Service charge capped at maximum of £5 per square foot per annum index linked	Preferred
Exclusion of liability for patent and inherent defects	Important
Repairing obligation limited by schedule of condition	Important
Landlord to provide full VAT indemnity or undertaking	Important

Occupiers often focus predominantly on the building requirements and only address the transaction requiremetns properly after the property has been identified. Negotiations commence during the site selection stage when the initial contact with landlords is made and if you are not fully prepared at this stage you will not be able to make the most of these discussions. Additionally it is essential to commence consideration of all issues as early as possible to ensure that by the negotiations stage the requirements have been fully developed, debated and confirmed.

As part of the planning process the occupier or advisers should undertake preliminary market research to assess availability and rental levels generally. The research should normally comprise:

- reviewing property journals and local newspapers;
- talking to local agents;
- identifying recent transactions;
- searching property databases;
- speaking to local government and public bodies, such as councils and chambers of commerce.

This general research assists in developing focused occupancy and transaction requirements.

The occupancy and transaction requirements are an essential element of the structured negotiation approach and are central to developing a strong negotiating position. As a result of having analysed and defined requirements in detail, the groundwork is laid for effective site selection and negotiation. By clarifying all preferences, the time and cost involved at each of the next stages is minimized.

Additionally, the process of defining strategy and occupancy and transaction requirements at the planning stage prompts debate prior to commencing site selection and negotiations. This ensures that options can be discussed and analysed properly and conflicts resolved before moving ahead with the transaction. At the end of the planning stage, the team is in a position to move into the next stage confidently with a clear view of what they wish to achieve.

2.2 SITE SELECTION

At the site selection stage the occupancy and transaction requirements are used to develop a shortlist of properties which best meet the strategic objectives and detailed requirements. As a result of the planning undertaken in the previous stage, the occupier is in a strong position to give clear guidance to the acquisition agent and the rest of the team to find and evaluate appropriate properties.

2.2.1 Property search

The first step is for the acquisition agent to undertake a market search. This usually comprises:

- circulation of details of the requirement to local agents;
- searches of agency databases;
- discussions with local agents.

In some cases other approaches may be relevant. For example:

- Contacting occupiers of target properties or those with similar business requirements to determine the availability of space within their portfolios. Sometimes client-to-client contact is the best route.
- Advertising the requirement in local newspapers, the property press or relevant trade journals.

The methods used should be tailored to the circumstances and the agent should be encouraged to consider innovative approaches where appropriate.

From the search the agent develops a shortlist. The client, agent and property adviser should then liaise to develop a shortlist of properties for inspection. The number will vary according to the nature of the project but inspection of approximately 10–15 properties is usually appropriate. The inspections should usually be undertaken by the client, agent and property adviser together. It is advisable that all are involved at this stage to ensure that the whole team can contribute fully to the process later.

It is advisable that as the client you take full notes on the inspections. It is best to prepare beforehand a checklist covering all the main issues to be noted. Figure 2.4 shows an example of a checklist for acquisition of an office in a multi-occupied building. This can be adapted for other types of property and the particular circumstances concerned. In all cases the checklist should be used only as a starting point for planning the inspection and you should consider fully what information you wish to obtain to ensure you achieve the maximum benefit from the time spent.

When conducting inspections it is advisable that as the client you avoid discussions with the landlord or his agents. Conversations during an inspection can often seem harmless but it is easy to be drawn into giving away information. Discussing anything with the landlord or their agent can prejudice the negotiations in two ways.

- You may release information which may weaken your negotiating position at this or a later stage. This can be as simple as revealing your time constraints, expressing an opinion about the quality of the space or comparing it with another shortlisted property. This kind of information may seem unimportant at the time of the inspection but may affect negotiations later.

'Here's a property you can grow into.'

- It can develop a line of communication between you and the landlord which may be used later to your disadvantage. It is important to avoid parallel negotiations between principals and agents. Direct approaches by the landlord to the occupier can put the landlord into a strong negotiating position by placing the occupier under pressure to agree terms quickly and without the support of advisers. The strongest negotiating stance is to keep all negotiations fronted by your agent. The agent is in a position to consider and advise on all options before agreeing any terms on your behalf.

Property:

Landlord
Letting agent
Managing agent
Other occupiers/neighbours

Occupancy cost:
   quoting rent
   estimated rent
   rates
   service charge
   VAT
   other charges

Location
External appearance
Age and condition
Entrance hall/reception/common parts

Facilities:
   toilets
   parking
   storage
   dining/meeting rooms
   other

Building management:
   building management system
   on-site management team
   security
   access

Space available
Layout and divisibility
Space inspected
Base building specification:
   ceilings
   floors and floor void
   decoration
   natural light and lighting
   air conditioning/heating
   glazing
   other

Comments/conclusions

Figure 2.4 Acquisition inspection checklist – office acquisition.

In general the approach should be to give away as little information as possible at all stages without appearing 'cagey'. Although it is important to maintain a good relationship with the landlord, becoming too friendly can be dangerous as it makes it hard to avoid giving away information and to take a tough stance in negotiations. At the viewing stage it is sometimes advisable to keep the name of the company confidential just providing limited information on the type of use and quality of the covenant. This deflects requests for information and discourages the formation of a direct relationship.

From the inspections it should be possible to narrow the choice down to a shortlist of four or five properties. It is important that the shortlist is no smaller than four as the next step is to undertake a full evaluation of the shortlisted properties and it is possible that out of this process one or more properties may prove unsuitable. There is often a temptation to focus exclusively on the preferred property but it is important not to put all your eggs in one basket. Competition will encourage all the landlords to make concessions and ultimately your negotiating strength is dependent on your ability to walk away from the negotiating table.

2.2.2 Property evaluation

This stage in the acquisition is crucial and rushing it should be avoided. It is essential that all the shortlisted properties are fully evaluated in respect of financial, occupational, and construction issues to achieve an effective ranking of the properties in order of preference. The aim of the evaluation process is twofold:

- to arrive at an overall ranking of the shortlisted properties to decide which to take forward into the negotiation stage;
- to construct a detailed assessment of all the relevant considerations to be addressed in the negotiations.

Crucial to the success of the acquisition process is a structured and rigorous analysis of all the issues that drive the decision. By evaluating the properties using a structured framework, you will be empowered to make choices not based on the size or colour of the entrance hall, but rather on the operational and financial issues that will determine the success of your business.

The property evaluation should be undertaken by you and the property adviser, agent and lawyer working together and it should address all business and occupational issues. The occupancy and transaction requirements will form the basis of the analysis but the review should also address other matters specific to the properties concerned.

The components of the review are as follows.

(a) Locational review

This assesses the advantages and disadvantages of the location and the extent to which it fits with occupancy requirements.

(b) Building review

This should address all the building specific issues addressed in the occupancy requirements including the following:

- Occupational considerations. This aspect should be assessed by the project manager and addresses all the construction issues relating to fit out and occupation. The objective is to assess the space for suitability for occupation and covers issues such as the quality of the base building specification, contributions to base building required from the landlord, suitability of the space for the proposed space layout and efficiency of the space usage. In general the aim is to identify any deficiencies in the specification, but equally where a building is overspecified there may be cost implications to be considered. The project manager and/or building surveyor will advise on the occupational issues.

- Repair and defects considerations. These issues are addressed in the building survey which should normally include both a fabric survey and a survey of the mechanical and electrical services (Section 2.1.1 above discusses instructing the building surveyor). The survey should identify **patent (visible) defects** and warn of potential **latent (invisible) defects** (sometimes known as latent defects, although both terms are sometime used with different meanings). Where potential inherent defects are identified, further investigation may be necessary. It should also assess the quality of design and construction and the condition of plant. Any wants of repair in the building generally and the demised area specifically will be identified.

  The types of problem that may be identified in the survey could include serious defects such as water penetration, timber decay, movement, use of deleterious materials in the construction (substances such as high alumina cement and asbestos), or structural design problems. More commonly the survey will identify problems with the condition of the building such as a need to repair particular items. Where serious defects are discovered you will have to decide, with the help of the professional team, whether the issues can be resolved as part of the transaction or whether you will need to eliminate the property from the shortlist. Repair problems can often be resolved by including relevant terms in the transaction.

- Environmental considerations. In some cases, you or the professional team may identify the possibility of contamination. In this situation it is important to determine what the risks are and the associated remedial costs. Further investigation will normally comprise an environmental audit, the cost of which can range from a couple of hundred pounds for a desk-based register search, to several thousand pounds for a full investigation with sampling and analysis. You should take the advice of the professional team in deciding what level of audit is needed.

  Any actual or suspected contamination problem will need to be taken into account in evaluating the property in terms of suitability for occupation and cost exposure. In view of the potentially large costs involved in remedying contamination, you will need to elevate the issue in the transaction negotiations to ensure the risks are mitigated. The issue is of greatest significance when you are acquiring a whole site, but it should not be ignored even where you are taking only a small part of a multi-occupied building as the cost exposure under a lease can still be very large. The terms you will need to consider are discussed in Sections 3.3.3 and 3.4.12. You will need to take the issue into account in negotiating the rental package or purchase price, as the value of your interest will normally be affected by a serious contamination problem or risk of such a problem.

An occupier encountering a contamination problem should ensure they take the advice of a suitably qualified professional. The acquisition team will be able to advise up to a point but you may require advice from a specialist lawyer or surveyor. Refer to a textbook on the subject for more detail. Laidler, Bryce and Boswall (1995) provides a practical summary and includes helpful advice on instructing advisers and transaction terms.

(c) Landlord review

The landlord's approach, management style and financial strength will have a significant effect on your occupation. Although some protection will be afforded by the legal documentation, it cannot cover every eventuality and cannot address the landlord/tenant relationship, which will colour the atmosphere in which you will operate on a day-to-day basis. While the nature of future landlords is outside your control (i.e. if the existing landlord sells his interest), you can assess the existing landlord and determine whether his approach will suit your occupational needs. Mutual trust and cooperation are fundamental to effective occupation of the premises. Landlords in financial distress, receivers and trader developers rarely intend participating in the form of commitment envisaged in the lease. If the landlord's objective is a quick profit you may decide that this is fundamentally in conflict with your objectives.

The landlord's financial strength can have an impact on his ability to perform his contractual obligations. Any doubt about this position needs to be addressed and resolved. The options are to withdraw from the property altogether, or to accept an element of risk but structure the transaction to minimize this.

In addition to evaluating the landlord's management style in terms of its effect on your occupation, you need to assess his approach to the transaction negotiations. You will need to decide how far you can trust the information provided by the landlord and extent to which you and the professional team will need to carry out due diligence.

Investigating and evaluating the landlord involves gathering as much information as possible from published sources and databases, and talking to people that know the company. Obtain their financial statements and corporate brochures and undertake a company search. Ensure that this information is analysed effectively, normally by an accountant. You need to understand exactly what motivates the landlord, and how he is likely to treat you once you are no longer a prospective tenant but instead a contractually bound stream of income!

Assessing the existing landlord does not obviate the need to protect yourself as much as possible in the documentation. The existing landlord can change his approach or may sell his interest. You should however satisfy yourself you will be starting off on a good footing.

(d) Transaction review

This is an assessment of the likely terms negotiable and a comparison with transaction requirements. You should take into account all the issues discussed in Chapter 3, in particular the term and options, rental package and any other elements of particular relevance to your business. Your professional team, and the agent in particular, will be able to make an assessment of the likely terms available and your negotiating strength for each property.

(e) Occupancy cost review

This is an analysis of the net present value of the occupancy cost over the expected period of occupation, taking into account all anticipated costs, including rental growth and inflation. It is important to factor in all property related costs, such as the fitting out and acquisition costs, and tax should be considered together with any tax savings, such as capital allowances, and local authority or other grants. Refer to Section 1.2 for a demonstration of this type of calculation.

In addition to the analysis of net present value you will need to consider the cash flow profile of each property. Where the landlord is prepared to make a capital contribution to fit out or to grant a long rent-free period, this may represent a cashflow advantage in comparison with properties with higher initial costs.

The transaction review and occupancy cost review are the two elements that will have to be developed and refined most over the site selection and negotiation stages. Your initial assessment of the terms negotiable and other costs will probably be inaccurate, and as you gather more information and develop negotiations with the landlord you will need to amend the assessment of terms and costs.

The five elements of the property evaluation are brought together in a comparative analysis for all the shortlisted properties which combines an analysis of all the qualitative and quantitative issues. Example 2.4 demonstrates the use of a weighted scoring system to examine and compare all the relevant issues.

Example 2.4   Property evaluation – comparative analysis

Issue	Weighting	Property A		Property B	
		Score	Total	Score	Total
Location:					
Proximity to staff, clients and suppliers	3	5	15	1	3
Amenities and facilities	3	4	12	2	6
Transport links	5	5	25	2	10
Building:					
Prominence/identity	1	3	3	2	2
Facilities	2	3	6	4	8
Size of available space	3	2	6	5	15
Flexibility of space for space planning	3	2	6	5	15
Specification	3	4	12	5	15
Fit-out costs	4	3	12	3	12
Repair and defects	4	4	16	3	12
Landlord:					
Credit risk	3	4	12	1	3
Management style	4	3	12	4	16
Willingness to negotiate	4	3	12	3	12
Transaction:					
Length of lease, break options etc.	4	4	16	3	12
Options to expand and contract	1	2	2	4	4
Occupancy cost:					
NPV analysis of projected occupancy cost over term taking into account the anticipated rental package, rent review structure, rental growth and inflation	5	2	10	5	25
Risk assessment – measure of the certainty associated with the occupancy cost projections	3	4	12	2	6
Total	–	57	189	54	176

In this fictitious example Property A has a central location close to transport links and facilities and the landlord is considered to be financially strong. Property B is much more economical but has disadvantages in terms of the location and the landlord. The unweighted scores are too close to assist with the decision, but the weighted scores indicate that Property B is the preferred choice. The weightings applied are based on the occupancy and transaction requirements developed in the planning stage but are defined in greater detail at this stage in response to the specific issues raised by the shortlisted buildings. This kind of comparative analysis has four important advantages.

- It is an effective way of defining explicitly the priorities that are fundamental to the decision. By explicitly attributing weightings both these priorities and the resulting score can be analysed and debated by all the decision makers involved.
- It assists in the assimilation and analysis of the information by enabling qualitative and quantitative issues to be assessed together.
- It produces a score which clearly justifies the decision to other decision makers in the organization. This can be important to ensure a confident, coordinated and harmonious approach during the remainder of the process. The importance of this should not be underestimated; a lack of consensus within the company and the acquisition team can severely damage negotiating strength.
- The explicit and detailed evaluation is a strong basis for ensuring that all issues are addressed properly in the negotiations.

Ideally all issues should be resolved as far as possible before embarking on negotiations. This maximizes your negotiating strength. By the time you are involved in negotiations, effort, time and money have been invested in the chosen property and there is usually a reluctance to withdraw from negotiations. It is therefore advisable to delay full-scale negotiations and any commitment to a particular property until all the shortlisted properties have been fully evaluated. This should include a preliminary technical appraisal of all the shortlisted properties to assess their physical suitability and general quality of construction. Ideally a full survey should be carried out for all the shortlisted properties but in most cases the costs of doing this are prohibitive and a survey can only undertaken for the preferred property.

The property evaluation may include a full due diligence exercise in which detailed information is assembled and checked. In many cases you will undertake the property review based on provisional information followed by a full due diligence exercise on the preferred property in the negotiations stage. In the documentation stage searches and pre-contract enquiries are undertaken by the lawyers, but this should be a confirmation exercise and it is dangerous to leave any significant information gathering to this stage by which time you will be fully committed to the property. It is important that any information that has formed the basis of the decision is confirmed before you commit to a binding contract. You should continually return to the comparative analysis and amend the figures to ensure that all new information is properly taken into account and assessed.

Once the comparative analysis has been completed the occupier can proceed to the negotiation stage. Negotiations will have commenced in the site selection stage and in many cases the first steps in the Heads of Terms negotiation described below would have been completed in parallel with the property evaluation. Having completed the evaluation, the occupier can now embark on full scale negotiations.

2.3 NEGOTIATIONS

This is not the forum for a detailed discussion on negotiating techniques but there are a few important points that should be highlighted. The business manager will normally be skilled in negotiations in his own field, but it is not uncommon for a manager who prides himself on his negotiating skills to leave them behind as soon as he steps out of his core business area and into the arena of occupational property. Conversely, landlords are on home territory when it comes to negotiating property transactions. They know exactly how to manipulate the situation to their advantage and occupiers all too often step happily into the traps.

Much of the advice given about negotiation generally applies to a property acquisition. One general principle is worth mentioning. A successful negotiation is one in which both parties' needs are satisfied. Your aim should be to reach an agreement that broadly meets the objectives of both parties and to achieve this you need, as far as possible, to maintain an amicable relationship. Even an antagonistic landlord will often respond favourably to efforts to be fair and reasonable.

Fundamental to this approach is a clear understanding of the landlord's perspective. It is only in this way that you will be able to structure a transaction that effectively meets both parties' needs. This involves investing some time in researching the landlord and talking to him and his advisers to get a full appreciation of what he wishes to achieve. You will have done this as part of the property evaluation and you can now apply that information to the transaction structuring process.

It goes without saying that it is also essential that you are clear about exactly what you wish to achieve. Without clear objectives in mind at all times during the negotiation, meeting those objectives will be haphazard. Having undertaken the planning work in the first two stages you should have developed clearly defined objectives and priorities. It is essential that you do not lose sight of these at any point in the negotiations.

Turning to the specifics of a property negotiation, even where you are not acquiring a property in an oversupplied market, in the initial stages of the negotiation you are in a strong negotiating position. At this stage the landlord will do all he can to court you as a prospective tenant and you will be free to pull out at any time. As the process continues your strength decreases. Even without urgent time constraints, the more time, money and effort invested in a negotiation the greater your reluctance to withdraw. By the time the Heads of Terms are agreed and you move into the legal drafting stage, it is hard to avoid a degree of commitment to the property, both psychologically and in terms of the resources invested, sometimes with the additional pressure of a looming deadline. This reversal in the power balance between the parties is illustrated in Figure 2.5.

The landlord's business is property. He understands the negotiation process. He is very aware of the reversal in the power relationship between the parties over the negotiation period and he will do everything he can to exploit it. It is essential that you confront this reality and structure your approach accordingly. The following two strategies will help.

● Finalize agreement in the early part of the process when your negotiating position is strongest.. Minimize the number of open issues left to the legal drafting stage. All agreements at the Heads of Terms stage should be clear and detailed. This requires the involvement of the whole team including the lawyer throughout the process. So often agreements thought to have been reached at the Heads of Terms stage are forfeited in the drafting stage because the agreement was ambiguous in the first place.

The legal document should be the fine print necessary to give effect to the agreement but no new substantial issues should be raised at that stage. Alarm bells should ring if you hear the words 'leave it to the lawyers'.

Detailed Heads of Terms are inevitably long and complex and you may feel you are 'over-egging' the process.You may meet opposition from your

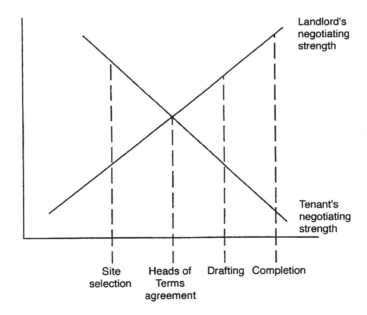

Figure 2.5 The negotiation relationship.

own side: senior management within your organization will question the mounting fees and your agent may advise that you are over-complicating the matter. If this happens, remind yourself of the reasons for the approach and reassure your colleagues that the time and money invested now should reap significant returns later on.

- Minimize commitment to the property throughout the process. At all points it is essential that you establish that you are still in control and are prepared to withdraw if there is any material erosion of the terms. An ability to minimize commitment to the property is dependent in large part on the time available and the alternative properties available. If possible, you should keep negotiating on other properties during the Heads of Terms stage. If you have followed the structured approach outlined above you will have left yourself enough time to progress negotiations on another property if negotiations on the preferred property break down. At every stage make sure you do everything you can to make it practical to withdraw from the negotiation.

In some cases the there may be pressure within the organization or the team mitigating against this strategy. Senior management may be keen to have the problem resolved quickly; an agent paid on a commission fee may be motivated to complete the transaction even if the terms agreed are not the most favourable. It is important to be aware of any reasons why any member of the team or the firm may be overly committed to the property and to bear this in mind when considering their advice.

Alarm bells should ring if you hear the words 'Leave it to the lawyers'.

It is tempting to relax as soon as Heads of Terms are agreed. You will have put a lot of energy into getting to this point and will wish to celebrate a successful negotiation. This puts the landlord in a strong position to erode the agreement you have worked so hard to achieve.

This is a crucial point in the negotiation as it sets the tone for the rest of the process. Some landlords will capitalize on weakness by a progressive process of challenging the agreement. It is only by remaining detached at this stage and avoiding a commitment to the property that you will be able to maintain a position of strength. At the first sign that the terms are being diluted it is essential to take a firm stance even if this means withdrawing from the transaction. By taking a strong approach at this point you will establish a position which will stand you in good stead for the rest of the negotiation.

It is important to retain this detached approach for the whole of the process. The professional team negotiating the legal drafting will have an unenviable task if the landlord's team senses an unwillingness to withdraw. It is essential that you are prepared to pull out at any sign of renegotiation, however late in the day. This does not apply to minor drafting points or differences of interpretation, but it does apply to any renegotiation of the material terms agreed at Heads of Terms stage.

If all else fails, if you have mistimed the negotiation and are under serious pressure to complete the transaction, you should hide this from the landlord. Never admit to serious time pressures or to a lack of available alternatives. The worst possible step is to take possession of the premises for fitting out prior to completing the formal document: this will make it almost impossible to withdraw and hence will tie your hands in the negotiation.

Having considered the negotiation relationship we can now turn to more general negotiating strategies. Throughout the process you should always ensure you retain control of the negotiations. Small tactics can give you an advantage. These may include the following.

- Hold meetings at your or your agent's office and produce the agenda and minutes.
- When attending meetings with your team, take regular breaks to discuss strategy.
- Most leases are structured with the tenant's obligations at the front and landlord's at the back; other clauses with significant impact for the tenant are also located at the end of the document. By the time you reach these parts in a drafting meeting your motivation to negotiate forcefully has diminished. A good tactic is to insist on starting with, say, the rent review clause, break clause or landlord's covenants section, so that these important issues are dealt with while you are fresh.

Advance preparation and being supported by a strong professional team will make a big difference psychologically as well as practically. You should always feel you are in control; as soon as you feel your control on the process slipping, step back, consider what is going wrong and address it.

If you employ all these tactics you will maximize your negotiating strength and should be in a position to negotiate a favourable transaction. In some cases the landlord may use tactics to undermine your negotiating strength. It is important that you are able to identify tactical moves designed to weaken your position and address these effectively. The following are some common examples of these.

- Direct approaches to the occupier. The landlord as a property professional will usually have an advantage if he negotiates with you directly. However informed or experienced you are in property acquisition and however good your negotiation skills, it is usually dangerous to negotiate directly. Multiple lines of communication mean that no one is in control of the negotiation. But most importantly, if you put yourself in the frontline you will be under pressure to make decisions without proper consideration and advice. The adviser or agent can always assert the need to take the client's instructions; you will not have the benefit of this excuse. You will find yourself agreeing points too quickly and revealing information that may prejudice your position.

  Your response to any direct communication from the landlord should always be to refer him immediately to the agent. In some situations, usually right at the end of the negotiation, it may be appropriate to hold a direct discussion to agree final outstanding points, but in the main you should avoid these.

- Undermining your professional team. If the landlord sees that you are being well protected by your professional team he may try to undermine them. This will usually involve implying they are being uncommercial or acting unprofessionally. You should usually interpret any attack on your professional advisers or on your overall approach as a likely sign that the team is serving you well. Unless you yourself have any reason to doubt their competence you should ignore the landlord's protests and encourage the team to continue. The last thing you should do is to allow the landlord to dictate that the professional team is bypassed or to let him undermine your faith in their advice.

- The non-negotiable clause. A commonly used tactic is 'the standard clause'. The landlord or his team will often be heard to argue that a clause is standard and non-negotiable. The implication is that you are being unreasonable in attempting to negotiate the point. It is a way of taking an aggressive stance while purporting to be reasonable. You are made to feel demanding and unreasonable.

  By agreeing this point, however unimportant, you accept the principle of non-negotiability and standardized clauses. This can have a significant effect on your negotiating strength on other terms. In some cases a landlord is genuinely not prepared to agree a term and it is a deal breaker. It is important that you distinguish these from the clauses that have simply grown up as standard terms because other tenants have agreed them. However commonly a term is agreed in other transactions, your concern should be to agree the best possible transaction taking into account the relative negotiating strength of the parties in this case.

Many standard terms developed in the landlord-dominated market of the 1980s in response to institutional requirements. It is a good idea to ask the landlord to explain the underlying reason for the clause. If he cannot justify a 'standard' clause convincingly it may be worth pressing the point.

That completes the general advice on negotiation strategy. As the client you should not normally take a front role in negotiations and all correspondence should be via your agent. However, it is still worth bearing in mind some of the legal issues that apply to the negotiation process. Firstly, it is important to ensure that you do not inadvertently commit to a binding contract in correspondence. Unless you wish to commit yourself, all correspondence in which you agree terms should be headed '**subject to contract**' or contain this wording within the text. Since the **Law of Property (Miscellaneous Provisions) Act 1989**, in most cases correspondence does not comprise a binding contract. However, there are some situations where a binding contract might be created and it is advisable that all correspondence during negotiations is made subject to contract. In some cases Heads of Terms are agreed subject to contract and survey. This additional caveat is not necessary but provides emphasis.

Another qualification sometimes added to correspondence is the phrase '**without prejudice**'. This is not usually applicable in acquisition negotiations as it should only be used where there is a dispute between the parties and the letter offers an agreement which the offerer does not wish to be binding in legal proceedings. This is discussed further in Section 4.2.1.

The methods adopted by the letting or sale agents are regulated by the criminal law and the common law of agency and contract, and more specifically by the **Estate Agents Act 1979** and **Property Misdescriptions Act 1991**. The law of estate agency, including the authority and duties of an agent, the information they can provide together with an explanation of types of agency instruction and marketing methods are discussed in Section 5.1.3. Please refer to this section as many of the issues are of relevance to the acquisition process.

You should consider whether you have any requirements in relation to **confidentiality**. It has become common for landlords to require that the terms of a transaction are kept confidential between the parties. In general this has resulted from the unfavourable terms agreed by landlords in the poor property market as landlords are reluctant to reveal details which may prejudice other negotiations. In some cases the confidentiality clause will only apply during the documentation stage and for a limited period afterwards, but in other cases it will apply permanently. Normally the clause is agreed as part of the Heads of Terms agreement but in some cases it may be appropriate to agree confidentiality at the start of negotiations.

The industry's Code of Practice on commercial property leases (Commercial Leases Group, 1995) discourages the use of confidentiality clauses. They have caused problems in the market as a result of the importance of market knowledge in the valuation process. Confidentiality has restricted the flow of information and caused difficulties in valuing, particularly for rent reviews where formal evidence of transactions often has to be provided. In some cases courts will override confidentiality clauses and subpoena the parties or their professionals to give evidence of transactions.

You should consider whether you wish to accede to a landlord's request for confidentiality. You may wish to publicize details of the deal if the terms are favourable and would represent good publicity for your company. In some cases you may yourself have a requirement for confidentiality and landlords may be agreeable in this case. Where there is no compelling reason for such an agreement, it is advisable to resist it. Where you do agree to confidentiality be sure that the clause does not apply to mandatory reporting (such as to comply with Stock Exchange regulations) or to alienation negotiations: if you wish to assign or sublet you will need to discuss the terms of your lease with the prospective tenant.

Having digested these legal issues and general negotiation strategies we can now turn to the specifics of the property negotiation process.

2.3.1 Heads of Terms

You should normally have at least three suitable properties available for negotiation. Any fewer could leave you without a fall-back property if negotiations on the preferred properties break down. Ideally four or five properties should be included on the shortlist for negotiation.

The negotiations will have commenced informally during the site selection process. The landlord will have quoted terms and may have provided a detailed proposal. Where he has made it clear he is not prepared to agree crucial transaction requirements the property should not normally have been shortlisted.

The first step in the negotiation process is to request from the landlord a **leasing proposal**. Essential requirements, such as the length of term and break options and other important terms should be specified. The landlord will have to assess where to pitch his proposal without knowing what you are likely to agree. He will want to make his offer attractive without limiting himself too much.

The landlord's proposal is the starting point for negotiations. The next step is to prepare a **counter proposal** detailing all the transaction and occupancy requirements including a rental proposal. This is the starting point for the negotiations and, although you need to be careful not to alienate the landlord by being too tough, you need to give yourself room for negotiation. The counter proposal should have input from the whole team. The lawyer will have an important input on all the legal issues and the project manager and building surveyor should advise on the construction issues. The proposal will reflect all the occupancy and transaction requirements formulated in the earlier stages and will include physical, technical, practical and financial issues.

Example 2.5 shows a fictional example of a counter proposal. In this case the premises are fitted out to a Category A standard (developer's finish) with the exception of certain deficiencies discovered in the construction evaluation. These deficiencies are addressed with a request for a number of landlord's contributions. The counter proposal includes a range of relatively complex terms, some of which will probably have to be conceded in the negotiations.

Example 2.5   Leasing counter proposal

A. Landlord Esq.   Subject to contract and survey
Property Estates Ltd
Main Street
Anytown

Dear Mr Landlord
Part first floor, Imperial House, No. 1 The High Street, Anytown

Thank you for your letter of 1 April. Our clients have now considered your proposal and I have been instructed to make the following counter proposal.

1. Tenant   Fine Business Ltd

2. Guarantor   Prosperous Company plc

3. Landlord   Property Estates Ltd

4. Guarantor   Rich Assurance plc (joint owner of Property Estates Ltd)

5. Demise   1000 sq m (10 764 sq ft)

6. Term   Ten years (within the security of tenure and compensation
               provisions of the Landlord and Tenant Act 1954)

7. Options   (a) an option to expand into the remaining space on the
               first floor on or before the fifth anniversary of the term on six months
               notice on the same terms excluding the rent-free period;
               (b)   a right of first refusal on the second floor on market terms;
               (c)   an unconditional option to determine the lease at the fifth anniver-
               sary of the term on six months notice;

8. Rent £10 per sq ft (£108 per sq m) based on the net internal area. Net internal area to be determined by an independent surveyor appointed jointly by the landlord and tenant on the basis of the Code of Measuring Practice (fourth edition).

9. Rent-free period   An initial rent-free period of 12 months to run from the date possession is granted, followed by an additional rent-free period of 12 months at the commencement of the sixth year if the tenant does not exercise the break option.

10. Rent review   A rent review to open market rent (upwards or downwards) at the fifth anniversary of the term. The rent is to be based on the determined net internal area.

11. Alienation   Assignment and subletting subject to landlord's consent (not to be unreasonably withheld or delayed). Assignments to any group company to be permitted without consent. There are to be no special restrictions on the nature of the assignee or subtenant and no guarantee is to be required of the tenant.

12. Service charge   The service charge is to be capped at an average of £5 per sq ft (£54 per sq m) per annum over the first five years of the term. An average cap is to apply in the second five years of the term to be calculated by reference to the Retail Price Index. Service charge contributions are to exclude the cost of improvements, and of remedying any base building defects and any costs recoverable under warranties or insurance.

13. Set off   The tenant is to have a right to set off the cost of remedying any landlord's breach of covenant against rent or any other payments due.

14. Repair   The tenant is to have an obligation for internal repair of the demised premises. This is to exclude structural repairs and the repair of any patent or inherent defects relating to defective design, workmanship or materials, or any damage arising as a result of such a defect. The landlord is to have an obligation to remedy such defects and is to provide a full indemnity in respect of any reasonable costs incurred where the tenant remedies such defects.

15. Alterations   Non-structural alterations to be allowed with landlord's consent (not to be unreasonably withheld or delayed). The tenant is to undertake fitting out of the premises to a specification to be approved by the landlord. The fit out is to include the installation of the items listed below under 'landlord's contributions'. The landlord may require reinstatement of any item included in the fit out specification, except for the items listed under 'landlord's contributions'.

16. Landlord's contributions   The landlord is to pay on completion of the lease or agreement for lease (whichever is earlier) the following allowances to the tenant in respect of base building fit out:

(a) The sum of £10 000 plus VAT to cover the cost of repairing the existing raised floor;

(b) The cost of installing floor boxes at the quantity of 1 box per 10 sq m (subject to the determination of the net internal area) at the rate of £80 plus VAT per floor box;

(c) A reasonable allowance of enhancements to lighting to meet current statutory regulations;

(d) A reasonable allowance to cover the cost of partitioning to separate the demised area from the remainder of the first floor.

17. Access   The landlord is to provide 24-hour access to the premises.

18. Landlord's services   The landlord is to have an obligation to repair and maintain the exterior, structure and all common parts and to provide services, including a manned reception and air conditioning, during the main business hours (8 AM to 7 PM Monday to Friday). The tenant can request that the landlord runs any service outside business hours, but the cost related to this additional service is to be reimbursed by the tenant.

19. Costs   The landlord and tenant are to be responsible for their own costs in agreeing and documenting the transaction.

I should be grateful if you could provide a response by close of business on Friday.

Yours sincerely

A.N. AGENT

The landlord will normally respond by agreeing some terms and rejecting others. The next step is for the agent to conduct negotiations with the landlord or his agent to try to reach agreement. The ranking exercise you have already undertaken at the planning stage will have established in advance your priorities and will guide you through the bargaining process. They will assist in ensuring that compromises are made effectively to reach the most favourable transaction.

You should avoid letting the landlord know which other properties are on your shortlist. This information will help him to assess your options and he may be able to find out what other landlords are prepared to agree. In all your negotiations you should ask the agents and landlords involved to treat the matter confidentially.

Once agreement has been reached on a significant number of issues it will be appropriate to meet with the landlord. Be aware however that attending a meeting gives a strong signal that you are keen. You should only agree to a meeting once you are close to agreement on all issues. It is usually advisable that at any meeting with the landlord you are joined by the agent and lawyer and, importantly, that you allow the agent or another adviser to conduct the meeting. If you try to take the lead you may enable the landlord to pressure you into agreement without proper consideration and you will open the way for him to make direct approaches later whenever it suits him.

It is central to the success of the meeting that you are prepared to end the meeting without reaching agreement if necessary. There will be a strong pressure to agree: a large team may be assembled on both sides and you will not want to return to your office empty handed. Do not let this pressure intimidate you into agreeing an unfavourable transaction.

Once terms are agreed these must be confirmed in detailed Heads of Terms. It is crucial that these are confirmed as they will form the basis of the legal drafting which translates the agreement into binding documentation.

Do not rely on the fact that at the Heads of Terms meeting everyone seems to know what is meant. People misunderstand and forget, or claim to. Ideally, the Heads of Terms agreed in a meeting should be put in writing and signed by both parties during the meeting. Once you leave the meeting with terms agreed the psychological commitment to the property is great. Subsequent confirmation can take days by which stage the lawyers and fit-out team have gained momentum and you have lost important time in other negotiations. You will be reluctant to take issue with small deviations from the terms you thought were agreed and the erosion of the agreement will be under way. Where you have negotiated a favourable transaction the landlord will be the one who benefits from any misunderstanding, so make sure you take control, clarify every point and confirm everything in writing immediately.

You may wish to agree at the Heads of Terms stage an **exclusivity period** for drafting, often known as a **lock-out agreement**. The landlord should be prepared to agree an adequate period for drafting during which he will not negotiate with other parties. This kind of agreement can be hard to enforce but it usually makes a big difference psychologically and the landlord will not normally negotiate openly with other parties. The disadvantage is that you will normally have to make the same commitment not to negotiate elsewhere, but as you will normally suspend other negotiations at this stage anyway this does not represent a substantial concession. Without an exclusivity period you will be at a disadvantage as you will be making a commitment to the property by investing time and resources in the negotiations and fit out, while the landlord may be negotiating elsewhere.

You should doubt the seriousness of any landlord who is not prepared to agree to a reasonable exclusivity period. Sometimes landlords will only agree to a very short period, say two weeks. This may be calculated to put your team under pressure and weaken your negotiating strength. You should avoid agreeing any period which would not give you time to negotiate comfortably. A month is normally a minimum.

Also to be addressed in the Heads of Terms is the issue of **legal costs**. Some years ago it was common for the tenant to agree to pay the landlord's costs in preparing and negotiating the documentation. More recently it has been usual for each party to bear their own costs. Generally, tenants' negotiating strength in the current market is such that you will not need to bear the landlord's costs and in any case you should be wary of agreeing to pay any of these costs as they are outside your control.

2.3.2 Legal documentation

Once terms are agreed and confirmed they have to be documented in a binding contract. The time involved to achieve this depends on the complexity of the transaction, the quality of the Heads of Terms (the extent to which there is need for further negotiation) and the efficiency of the lawyers on both sides. However efficient and willing your solicitors, their hands will be tied until the landlord's solicitor provides draft documentation and title documents.

Where the occupier is taking a lease it is normal for the drafting negotiations to commence with preparation by the landlord of a draft lease and other appropriate documents. Bear in mind the negotiating advantage that this gives the landlord: you and your lawyers will inevitably be on the defensive trying to amend the landlord's solicitors' drafting. You will gain a significant advantage by deviating from conventional procedure and insisting in the Heads of Terms discussions that your solicitors prepare the first draft, but it will probably be difficult to persuade the landlord to agree to this. If the landlord insists on preparing the draft, be sure to take a forceful stance on the document and not to be intimidated by the number of amendments for which you will have to argue.

Your solicitor, agent and adviser will review the draft documents and discuss the problems and issues with you. The draft should reflect all the Heads of Terms agreed. If there is a significant deviation from the agreed terms your only option is to threaten to pull out unless the documents are redrafted; any other approach will significantly weaken your negotiating position as discussed above.

Normally, your lawyer will send an amended 'travelling draft' of the documents back to the landlord's lawyer incorporating all the relevant comments and suggested amendments. You may be able to let the lawyers debate the technical drafting without you. They should keep you informed of the state of negotiations and the outstanding points. At a later stage, depending on the extent of disagreement and the time deadline, the parties and their advisers can meet to agree the documentation.

Refer to Chapter 3 where the drafting issues are explored in detail.

Part of the legal process may involve your solicitors checking the landlord's or vendor's title and carrying out pre-contract enquiries and searches. Good title must normally be shown to establish that the landlord or vendor is legally empowered to grant the lease or sell as well as to ensure you will be able you to operate your business successfully. You should note that where you are taking a lease, any restrictions on use and occupation included in the landlord's title are likely to be enforceable against you as occupier, even if they are not expressly contained in your lease. If your solicitors are not checking the title, as is sometimes the case for short leases, it is important that they obtain details of any restrictions on use and occupation from the landlord's solicitors. Also, if the landlord's interest is mortgaged, the mortgagees will probably need to consent to the grant of a lease.

It is a standard provision of a contract or lease that the purchaser or tenant cannot rely upon any representations made except those made in writing by their lawyers to yours. Unless your lawyer can delete or amend that provision it is vital that any information you are relying upon given to you directly or via your agents is confirmed via the lawyers so as to establish possible avenues for redress if the information proves inaccurate.

The key matters the lawyer will have to attend to at the documentation stage are:

- Preliminary enquiries of the vendor/landlord. These should reveal practical information, such as service charge information or disputes. It is not unknown for very little information to be disclosed in replies. You should note the following.
  - Your lawyer should press for satisfactory replies where inadequate replies (or worse, none at all) have been given.
  - Given the standard contractual provision whereby you can only rely upon these replies, you should ensure that if any representations have been made to you or your agents that you intend to rely upon, confirmation of the information is requested in the preliminary enquiries.
- A local authority search. This reveals information gathered from several departments in the local authority and will reveal matters such as whether the roads required for access are maintained at the public expense, whether the premises have the benefit of mains drainage and whether any development at the premises is authorized by a subsisting planning permission or (if not) whether any enforcement notice has been served.
- Other searches of relevant authorities, for example a mining search or commons search, depending on the location of the land or premises.

Much of the information extracted from the replies and searches will be of practical use to you, for example in confirming the previous and budgeted service charge, other outgoings, maintenance contracts and other building management issues. If you have queries which you would like your solicitors to raise, you should tell them at an early stage.

For both a lease or a purchase the transaction may be completed in one stage or there may be an exchange (agreement for lease or sale) followed by completion. It often comes as a surprise to clients but exchange and completion rarely require your attendance at a meeting. They can usually be carried out on the telephone between the lawyers and effected by the giving of professional undertakings. A deposit may be payable at exchange, particularly for a purchase. The deposit is normally non-refundable but this depends on the terms of the contract and the other circumstances.

If money is changing hands at completion your lawyer will prepare a completion statement and you will need to arrange funds, although these may be coming from a lending source with whom your lawyer may liaise directly. After completion he will attend to payment of any stamp duty and any requirements for registration at the Land Registry, for which a fee will be payable, based on a scale according to value and/or rent.

Once the transaction is agreed, but prior to exchange or completion, your solicitor should prepare a report detailing the terms agreed and their implications. It is important that you review this properly and consider the transaction and all the terms. If you think that the transaction or any part of it does not meet your business needs you should seriously consider going back to the negotiating table or pulling out altogether. This is your last chance to decide whether to proceed with a transaction that may affect your business for many years. Do not treat it lightly.

At this stage you should also ask your professional team to consider the transaction in a detached way and to advise you whether they consider you should proceed. Many of the issues are commercial decisions that may not fall within their remit, but they will have extensive experience of transactions and their views will be helpful to you in making this final assessment.

If you decide to proceed you may need to fit out the premises. In practice, planning of the fit-out process commences prior to completion of the documentation and runs parallel to the transaction negotiations, but the process cannot commence in earnest until there is a signed document in place. In the next section we take a look at the fit-out process.

2.4 FIT OUT

The structure of the construction team is outlined above in Section 2.1.1. The team has a role both at the site selection and negotiations stage in advising on construction aspects of the property and, where there is a need to undertake works to fit out the premises, during the construction project.

This section envisages a situation where the premises are fitted out to a developer's finish and the occupier needs to undertake fitting out (Category B) works together with some enhancements to the base building (Category A). The landlord may be making contributions to the cost of Category A works. Refer to Section 3.4.7 for an explanation of these definitions.

The construction works may not be of this type. For example, the premises may already be fitted out and the occupier only need to undertake minor alterations to meet their needs. Conversely, where the occupier takes a pre-let or purchases a freehold on a building not yet constructed, there may be more extensive technical issues to be addressed. The information contained in this section should have applications in most construction situations, but will have to be adapted to the particular situation. In all cases, you will need to take the advice of your professional team who can advise you on the approaches most appropriate in the circumstances.

2.4.1 Project management

The following sections explain the process of planning and managing a fit-out contract. The initial stages of the fit out project will run in tandem with the acquisition process and in particular will feed into the planning and site selection processes described in Sections 2.1 and 2.2 above.

The fit-out process depends on the nature of the project and the procurement route adopted (Section 2.4.2 below). The process described here will apply in most cases, but will have to be adapted to the circumstances and, in particular, where a non-traditional route is used, such as design and build, the process will have to be modified. However, in all cases, regardless of the procurement method, all stages need to be addressed, although the order may be different and they may be carried out by different members of the team or contractor. Figure 2.6 illustrates the process.

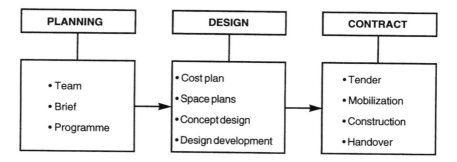

Figure 2.6 The fit-out process (traditional contracting route).

(a) Planning

The first step is to identify the **in-house representative** or management team who will represent the occupier throughout the fit-out process. Representation by committee or subcommittee can be time consuming and lead to indecision; a committee of one supported by the partnership or directors is often most effective. The in-house representative may be the process manager acting on the acquisition.

The in-house representative needs to have adequate time to carry out the role properly. All too often it is undertaken by an individual who is expected to carry out a normal full-time workload in addition to representing the needs of an organization in what is likely to be an important and time-consuming project.

It is equally important for the organization to identify the role of the in-house representative so that he is clear about his remit. The in-house representative must be of sufficient seniority in the organization to have the power to give the necessary approvals throughout the life of the project or have quick access to those who are able to sign off on key decisions. If it is necessary for certain decisions to be referred to regular board meetings, the meeting dates should be identified.

The **professional team** is discussed in Section 2.1.1 above in the context of the acquisition as a whole. The construction team shown is simplified and in practice other professionals may be required. The project manager should identify the additional professional assistance which is required such as quantity surveyor (cost consultant), interior designer, architect, space planner, environmental services engineer, structural engineer, acoustic engineer, catering consultant or lighting consultant. A **planning superviser**, often the project manager, will be required from the planning stage in order to comply with the Construction (Design and Management) Regulations 1994 – see Section 4.10.2. Once you have established which professionals are required, the project manager can proceed with competitively procuring their services based upon a clearly defined schedule of duties.

The project manager should identify a clear and effective project reporting structure and procedures setting out the communication routes within the professional team and between the team and the employer.

The project manager must work with the occupier in developing the **brief**. An overview brief must be set out identifying the main objectives of the project, setting the standard and image of the fitting out and identifying any programme and budgetary benchmarks which must be met. It is essential at this stage that long-term space requirements are identified. The in-house representative will be expected to identify the key in-house department heads who can advise on their requirements and preferences. This information is then translated into a briefing document which should be signed off by each department head. More detailed briefings with special groups such as the information technology department will continue through the early stages of the project. Once established, the detailed short- and long-term space requirements are factored into the design and cost plan.

A schedule which identifies all the components which must be organized to complete the move process should be prepared. Typically this is prepared by the project manager who identifies who will take responsibility for the organization of each of the activities involved in the project including construction and non-construction work such as removals, artwork and planting, maintenance contracts etc. Quite often the employer has the in-house resources and ability to organize some of the non-construction activities, especially in large organizations with property or facilities management departments.

At the earliest date, normally once the initial brief has been taken, the project manager should prepare a **master programme** identifying key activities and sign offs. Once this programme has been approved it is the project manager's role to drive the project to meet these key dates and advise the consequences in terms of cost and quality of product, in advance, if there is a prospect of them not being met.

(b) Design

It is essential that an outline **cost plan** is prepared in the early stages of the project. Initially, a concept cost plan should be developed based upon the cost consultant's previous project knowledge and experience. This will be developed and refined once space plans have been prepared and the designers have developed their concept design. It is important to identify what is excluded so that appropriate allowances may be made against these items.

In many cases the first cost plan will not meet your requirements and will require explanation and cost options to be identified. The analysis of objectives and requirements in the planning stages of the acquisition will stand you in good stead for the decision making at this stage. Further financial analysis to fully evaluate the detailed options will assist in making the right choices to meet cost and other requirements.

Analysing and debating the cost plan should start during the planning stage of the acquisition process discussed in Section 2.1 above. It should be based on the overall business plan and strategic objectives and should be used to drive the space planning and design process. This ensures that the project is driven by business objectives and not by design considerations.

Once cost and other requirements have been established, the designers will translate these into space requirements based upon a headcount and typical workstation footprints for each user type. These details will be worked up into **block plans** and **space plans** which test whether and how the organization can fit into the building. On a small project block plans may not be needed. The in-house representative will be expected to sign off these details to enable detailed design development to proceed.

The project manager must ensure that the designers develop the concept design within the parameters of the brief. **Concept design** presentations, in particular for the interior design, should be made to senior management to ensure that the design is developing in the correct direction. These presentations are at one of the most important stages in the design process and require key user feedback which may culminate in design concepts being altered or rejected.

The cost plan, space plans and concept design should be signed off at the same time. The project manager, in conjunction with the designers and the in-house team, must ensure that the sign off dates for concept design are met, to enable the design to proceed to the detailed stage.

Once the space plans, concept design, cost plan and project programme are signed off the designers will commence their detailed design work. The project manager's role is to ensure that the designers develop the concepts within the cost parameters, to advise the occupier where these are not being met and to advise whether value for money is being achieved. The project manager's regular progress reports should include a cost appraisal which will evaluate the financial impact of the design development. The project manager will establish a sign-off procedure to ensure that the key programme dates are met.

It is unlikely that any employer will move through the design, procurement and construction process without requiring design changes to be made. Changes are often essential, but you should bear in mind that most changes will incur penalties in terms of cost, time and/or quality. A risk assessment should be undertaken when considering the effect of any change. Once the construction works commence on site the effect of changes becomes more significant.

(c) Contract

The next stage is to go out to **tender** to contractors for the construction work. Once the procurement route has been selected, the design team should develop a list of suitable contractors. Contractors should then be shortlisted based on their ability to meet quality, time and cost constraints, their previous experience, financial stability, the size of the organization, flexibility and current workload and resources. The professional team will normally have experience dealing with a number of the contractors and will have a good appreciation of the reputation of others. It is advisable that references are taken and a financial check made on all shortlisted contractors.

Tender documents must be prepared setting out the project details and constraints, such as access arrangements and occupational information in addition to details of the works and workmanship requirements. These documents are typically prepared by the project manager who brings together the design information into the tender document. Once tenders are returned the project manager must analyse and compare the submissions in order to recommend which contractor is offering the best value for money.

Having selected one or more contractor, the project manager can prepare the contract document. In complex cases the assistance of a lawyer may be required. Once the contract is completed, the contractor can mobilize his workforce and place subcontracts and orders for materials. This process usually takes about four weeks although this depends on the size and nature of the project and the availability of supplies. In very simple cases, the **mobilization period** can be as short as a few days. It is not advisable to commence work on site prematurely as delays once on site can result in non-productive costs and can damage the morale of the workforce.

Once on site, the project manager's role is to coordinate the completion of detailed design production in accordance with the contractor's **construction programme** to enable orders to be placed on time. The project manager monitors the progress of the works against the construction and master programmes. It is essential that the project manager prepares regular project progress reports which, if required, can be tabled at board level.

The **quality** of the installations should be constantly evaluated by the design consultants. In the case of the engineering services these checks culminate in witnessing tests for certain parts of the installations such as the electrical power circuits. Monitoring and testing quality is crucial to the success of the project as cost and time pressures can mean that quality is prejudiced.

It is advisable to benchmark the cost and standard of the fitting out installations against other similar projects. Normally the professional consultants can identify similar installations from their experience.

Regular **payments** to the contractor(s), normally monthly in arrears, should be agreed with the project manager who is also responsible for agreeing the overall final account for the works. The project manager coordinates the **handover procedure** which includes **snagging** (identifying any defective installations), arranging for the appropriate training and demonstrations, ensuring the operating and maintenance manuals are completed. After completion the project manager ensures that the defects that have been identified and other defects which become apparent during the pre-specified defects liability period are made good. The handover procedure should clearly identify when the responsibility for the premises and installations passes back to the employer.

2.4.2 Procurement

If you are placing a fit-out contract you will need to decide between a number of procurement routes and pricing methods. This section gives a brief description of the main issues to enable you to take an active role, alongside the construction team, in placing the contract.

Your objectives as an employer of construction professionals and contractors will normally be to complete the project within cost and time constraints to an acceptable level of quality. The objectives of time, cost and quality often mitigate against each other and you will need to decide your priorities as a basis for deciding on the best contracting method. For example, if you are under particular time pressures as a result of having to vacate your existing premises, completion of the project by the deadline may be your most important objective and you may be prepared to risk cost overruns to avoid programming delays.

In most cases contractors wish to develop and maintain good relationships with their customers and will aim to quote a high enough price to ensure that cost, quality and time requirements can be achieved. However, at the bidding stage contractors can sometimes be motivated to underbid in order to obtain the contract with the aim of finding ways later of passing costs back to you or justifying time extensions by problems caused by you or a third party. When you are agreeing the contract you should aim to tie down issues as clearly as possible to minimize the scope for overruns and disputes later. This is partly dependent on developing clear objectives in the early stages, translating these into well-drafted tender documents, and not changing the requirements significantly later on in the project. The project manager can assist in developing requirements and deciding between the competing objectives of time, cost and quality.

Leaving enough time is crucial to the success of the project. You will need to determine the type of construction contract, pricing method and contractor most suitable for the project. You also need to invest time at the contract stage to ensure you are properly protected. Leaving gaps in the agreement is a recipe for dispute at a later stage and will give the contractor the opportunity to wriggle out of his obligations.

The following sections provide an outline of the main issues relating to construction contracts. You should take the advice of a building surveyor on the procurement route and use of standard form contracts, and if necessary a lawyer can advise on the terms of the contract if the standard form is to be modified.

(a) Types of construction procurement

The four main types of construction contract are those described below. There are other types of procurement, such as prime cost contracting, but these are less commonly used. In most cases an occupier undertaking fit-out works will use one of the first two options, either the traditional contract or the design and build method. For larger projects one of the other two approaches may be appropriate.

Traditional contract

The general scheme under a traditional contracting arrangement involves the employer engaging his own team of consultants and professionals to design the works to prepare the necessary contract documents (Figure 2.7). The main contractor is provided with drawings and bills of quantities or a specification in order for him to prepare his tender, and then enter into the building contract with the employer.

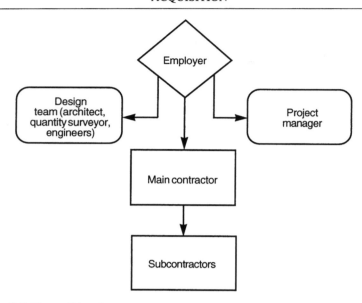

Figure 2.7 The traditional contract.

A standard form of contract will normally be chosen from one of those published by the Joint Contracts Tribunal (JCT). Other standard forms are used less commonly. The JCT forms contain provisions regulating the rights, duties and obligations of the parties, the powers and duties of the professionals, procedures for resolving disputes. There may be financial consequences specified in the contract where the employer or professional team do not comply with their obligations, for example where there is delay in provision of drawings. The contractor will normally be responsible for the performance of the subcontractors, except where subcontractors are nominated by the design team. The contractor will normally have to pay liquidated and ascertained damages (a specified amount payable for each week late) where the work is not completed according to programme, unless the employer, professional team or a third party is to blame.

Design and build contract
In a design and build contract (Figure 2.8) a complete service is provided, and the contractor is responsible for both design and construction. This approach can have advantages as the employer only has to deal with one party who has liability for all aspects of the project, and the approach can avoid some of the coordination problems that sometimes beset traditional contracting arrangements. However, in practice the employer may need to obtain separate contractual agreements with the contractor's professional team in order to ensure they have a liability. It is often advisable to appoint a separate consultant or agent,

known as the 'employer's agent' or 'representative' usually a project manager, acting on behalf of the employer, to prepare a brief, evaluate the contractor's proposals and monitor performance. As a result of the total reliance on one contractor for all aspects of the project, the design and build contract is only as good as the particular contractor used and you need to be fully satisfied of his ability to perform.

In some cases this route can be the most cost effective. However, the occupier should beware the contractor who quotes attractive figures at the bidding stage in the knowledge they can claw back losses during the project. Without the protection of a project manager acting as employer's agent, the design and build route can result in the occupier being poorly advised and exposed to cost and time overruns or problems with the quality of design or construction. It is important that a detailed brief is established setting quality standards, and the scope of work required.

The JCT provides a standard form of contract for the design and build route. It provides that the contractor is required to design in accordance with the usual obligation for professional services of exercising reasonable skill and care. It is sometimes possible to negotiate a higher level of responsibility, such as 'fit for purpose'.

You will normally need to investigate the design and build contractor's financial position and ensure there is adequate professional indemnity cover.

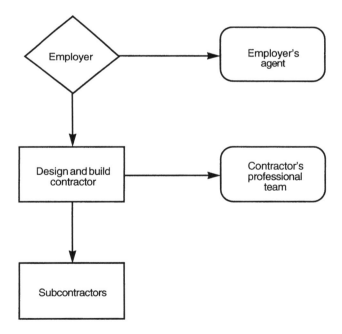

Figure 2.8 The design and build contract.

You will need to be sure that if it comes to a claim against the contractor, and he is left with a residual liability after subcontractors and professionals have been pursued, he has the financial means to support the claim.

Management contract

The management contract (Figure 2.9) follows broadly the same structure as the traditional approach. The difference lies in the terms of the contract which pass more of the risk onto the employer, and aim to create a relationship of co-operation between the employer, contractor and professional team. The approach is to avoid the adversarial relationship that can develop in a traditional contracting structure by making the management contractor work closely with the professional team, coordinating and supervising the works contractors but also assisting in the preparation of the design programme and giving cost advice. Payment is usually on the basis of prime cost (the actual cost of the works) together with a fee calculated as a percentage of the prime cost.

One of the main problems with this approach is that the prime cost and the fee cannot be accurately assessed when the contractor is appointed. There are additional risks resulting from the fact that the management contractor will only be liable for cost and time overruns to the extent that he is negligent. This is because the management contractor is usually only liable for the default of works contractors to the extent he can recover from them. You will not usually

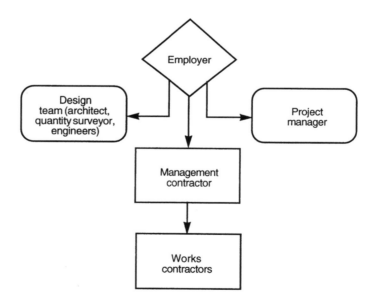

Figure 2.9 The management contract.

have a completion or cost guarantee from the contractor and will have to rely on recovery from numerous works contractors.

Management contracts are generally only appropriate for large projects where 'fast tracking' is required. It is possible to instruct the contractor early in the pre-construction period, because the need to finalize documentation prior to contract is eliminated, and thus the design and construction periods can be over-lapped. This is one of the main advantages of this route. On the other hand, the employer is exposed to significant risk under this arrangement in terms of time and cost.

The JCT Standard Form of Management Contract 1987 is the main standard form used.

Construction management

Construction management (Figure 2.10) is a development of the management contracting route. Here, the construction manager is a member of the professional team combining the project management and construction management functions, and the employer has a direct contractual relationship with the trade contractors undertaking the work. The construction manager has overall responsibility for coordinating the contractors but is not liable for a trade contractor's breach of contract.

The main advantages of this approach are the control over trade contractors afforded to the employer, the scope for fast tracking and the possibility of avoiding adversarial relationships as a result of the construction manager being on the professional team. The problem is that all the risk is left with the employer and his recourse against a defaulting works contractor will be limited to the scope of the individual contract. In the event of a problem the employer is faced with a large number of parties who may be liable. It is generally only suitable for use by experienced procurers of construction works.

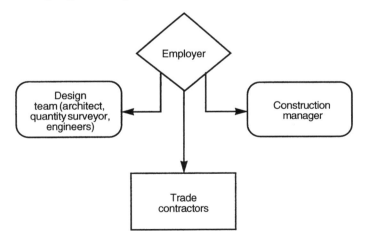

Figure 2.10 Construction management.

Standard forms are not yet published for this type of contract.

(b) Pricing methods

- Fixed price. This is a common pricing method and imposes an obligation on a contractor to complete the works for a fixed price. This has obvious attractions from an employer's standpoint and equally obvious disadvantages from that of the contractor. The work may prove more difficult than anticipated, and cost increases may affect the price of materials and labour. However, very few fixed price contracts are in reality what their name might otherwise suggest. The standard form of fixed price building contract most commonly used (JCT Standard Form of Building Contract Private with Quantities, 1980 edition) sets out a number of grounds upon which a contractor is entitled to claim extra payment. Alternative provisions for fluctuations are also included, to be selected by the parties. Some items of work will be covered by provisional sums and adjustment can be made for variations and for some instances of delay and prolongation.

  If loss and expense claims are to be kept to a minimum, the works need to be sufficiently designed at tender stage in order that bills of quantities can be prepared to enable the contractor to calculate a price. Where such advance design is not possible (for example because an early start on site is required), a fixed price contract is rarely the most suitable, as it would be likely to give rise to a large number of claims.

- Re-measurement. Under a re-measurement contract, a contractor is entitled to be paid for all works properly executed (be it greater or less than originally envisaged) and no more than an indication of the likely cost is given to the employer at tender stage. This is usually based upon approximate or estimated quantities or schedules of rates prepared by the employer's quantity surveyor. The actual price payable is normally determined by measurement and valuation at the rates shown in the bill of quantities or schedules of rates. An example of this form of contract is the JCT Standard Form of Building Contract with Approximate Quantities, 1980 edition.

- Cost reimbursement. The contractor is paid the prime cost of labour, materials, plant and subcontractors in addition to an amount for overheads and profit. The four common forms of this method are:
  - cost plus percentage fee
  - cost plus fixed fee
  - cost plus fluctuating fee
  - target cost.

The JCT Prime Cost Contract, 1992 edition is an example of the first two of these. Under this form, the fee can either be a fixed sum of money or a percentage of the amount of the prime cost certified by the architect.

● Provisional sums. These are sums of money included in the contract by the employer each relating to an item of work which cannot be detailed accurately for tender purposes. At the time of preparation of the final account, the provisional sum is deducted from the contract sum and the work carried out is valued in accordance with the contract provisions relating to valuation of variations. It is by no means unusual for 'lump sums' to be little more than an aggregate of provisional sums. Where there are a large number of provisional sums in the contract price the final price is likely to differ significantly from the estimate.

(c) Construction insurance

Various forms of insurance will be taken out in respect of a building contract, and these will cover two distinct areas – liability and property. The liability insurance will generally fall into two parts, that for death or injury to third parties, and that for damage to third party property. The property insurance, generally known as **'contractors' all risks'**, will cover the works together with unfixed goods and materials. The JCT standard forms of contract provide that this insurance may be taken out by the employer or the contractor, although in practice it is usually placed by the contractor. Where the works relate to part of a multi-occupied building the landlord may place the cover. In this case the contractor will normally require an express subrogation waiver from the landlord's insurers which prevents the insurer pursuing the contractor for any damage. In some cases a policy in the joint names of either the contractor and employer or contractor and landlord is appropriate.

(d) Collateral warranties

Where you are undertaking substantial works it may be appropriate to ensure that collateral warranties are available to those with an interest in the property who would not otherwise have a direct contractual relationship with the contractors and professionals. These enable the liability that the contractor, supplier or professional has to the employer to be extended to other parties such as lenders, purchasers, and future tenants. In some cases, without such warranties you may encounter difficulty if you wish to dispose of your interest in the property. The warranties are normally only valid for a limited period (usually 12 years) and usually include a number of exclusions which can mean they are of limited value. You should take legal advice on the need to obtain warranties in a particular situation.

(e) Performance bonds

For larger fit-out contracts, it may be necessary to obtain from the contractor's bank, insurer or other financial institution a performance bond which provides funds in the event the contractor fails to complete the contract, particularly in the case of insolvency. The value of the bond is commonly approximately 10% of the value of the contract, designed to cover the additional cost that would be required to complete the works. The contractor has to pay a fee to the financial institution, which depends on the credit status of the contractor, typically 1–3% of the value of the bond. It will represent an additional cost to the contractor and will probably increase the price charged by them. The professional team can advise on whether a bond is necessary.

FURTHER READING

Commercial Leases Group (1995) *Commercial Property Leases in England and Wales: Code of Practice,* RICS Books, London.

Laidler, D. W., Bryce, A. J. and Boswall, J. R. G. (1995) *Guidance on the Sale and Transfer of Land which may be Affected by Contamination*, Construction Industry Research and Information Association (CIRIA), London.

Lock, D. (1992) *Project Management,* Gower Publishing, London.

Transaction structure and terms | 3

KEY POINTS

- The types of ownership structure you should consider
- Understanding the principal terms of the transaction
- Assessing the implications of common lease terms and the drafting options available to protect you in the future

The transaction requirements discussed in Chapter 2 need to be based on a full understanding of the legal and technical framework and appreciation of the implications of all terms. By understanding fully the choices available, the implications of each option and the costs, you will be in a position to develop an effective set of transaction requirements as a basis for the negotiations. This will enable you to maximize your negotiating strength and achieve the best possible transaction.

This chapter touches on a range of transaction structures with a focus on leasing of the whole or part of a new or second-hand building. This is the area where many occupiers struggle because of the legal and technical complexity and the need for informed decision making. There are several alternatives to the basic leasing structure including purchase of a freehold or long leasehold interest, development of a bespoke building on a pre-let or purchase arrangement, finance and equity leases and other funding related structures. Reference will be made to these types of alternatives but the focus is on the terms of a leasing structure. Much of the advice and information contained in the chapter relating to this route can be adapted where alternative options are chosen.

Where the occupier takes an assignment of an existing lease there will not normally be any scope to renegotiate terms, unless the landlord particularly wants you as an occupier, but you may be able to reflect onerous terms in the transaction negotiated with the outgoing tenant. Where you are taking an assignment situation this chapter will help the occupier to evaluate the lease terms imposed.

3.1 OWNERSHIP STRUCTURE

This section reviews the broad ownership options available and the issues to be considered when deciding between the various purchase and leasing structures. The main ownership options are shown in Figure 3.1.

3.1.1 Purchase

In some cases the purchase of a freehold interest or long leasehold interest may be appropriate. A long leasehold is normally defined as a lease for a term longer than 25 years without regular rent reviews.

The company's cashflow position and the availability of finance will determine whether the purchasing option is practical. Where it is a practical option the lease/purchase decision should be based on a broad range of occupational, strategic, financial and legal considerations. A full quantitative and qualitative analysis of the options including cashflow (net present value) analysis, risk analysis, and a review of the accounting and business impact (Section 1.2) will assist in the decision-making process.

The following are some of the issues that will affect the decision.

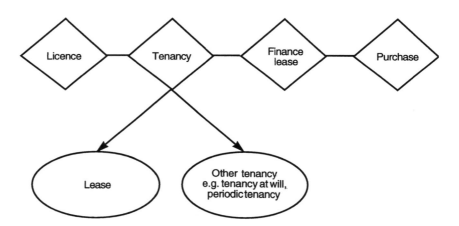

Figure 3.1 Ownership options.

- Cost comparison of the leasing and purchase options. The attractiveness of purchasing, compared with leasing, will depend partly on the relationship between the cost of capital and market yields applied to property investments. As a general rule, purchase is attractive at times of low interest rates and high property yields. Where the yield reflected in the capital price for the property is higher than the occupier's cost of capital the rent payable under a lease will be higher than the cost of financing the purchase.

  Another general principle is that purchase will be attractive at times of high inflation, when a fixed occupancy cost based on the depreciated purchase price plus financing costs will be preferable to the alternative of rising rental costs.

- Investment analysis and funding. The access to funds will affect the ability of the firm to finance a purchase. Assuming that there is a limited supply of funds, the investment of capital in property diverts funds from the firm's core business and from other investments.

  In addition to the actual costs of ownership, the opportunity cost of investing funds needs to be considered. The analysis should scrutinize carefully the projected return from the property investment compared with the return on alternative investments, including the core activity of the business, together with the risks associated with each.

  The company must be satisfied that it has the management and financial resources to maximize the return on the investment. If the company is unable to approximate the return that a property company would achieve it is unlikely that the purchase option will be advantageous.

- Risk. The potential for capital gain can be attractive but it should be assessed in the context of the risk of capital loss. Other risks also need to be taken into account, such as the risk of a landlord not complying with his obligations, the risk of capital expenditure, for example on repairs and defects, and the different problems associated with disposal of leases and freeholds. Sensitivity analysis, or, where possible, Monte Carlo analysis will assess the impact of changes in significant variables such as the rental growth rate, capital growth rate, cost of capital and tax rates (Section 1.2).

- Tax issues. The tax implications of the leasing and purchase options should be analysed fully and factored into the net present value analysis. The taxation of any capital gains on disposal of an investment property, the availability of capital allowances, deductibility of rent from income profits and VAT costs are some of the main tax issues to be considered (refer to Section 1.6 for details).

- Accounting impact. Even if an option is unattractive in economic terms, the accounting impact may make it worthwhile. Equally, where the cost of capital is lower than the property yield a purchase may be attractive in economic terms but the additional depreciation may make the investment unattractive in accounting terms. In general, a purchase will bolster the balance sheet, will be damaging to the profit and loss account in the short run and will benefit the profit and loss in the longer term. (Refer to Section 1.5 for details.)
- Market availability. In some cases the lease/purchase decision may be imposed by the availability of space. In some cases the ownership structure may be dictated by the vendor or landlord and if the property meets the occupancy and transaction requirements in other respects you may be prepared to accept this transaction structure.

  If you are unable to agree lease terms on an acceptable basis it may be preferable to move to a purchase transaction.
- Flexibility and control. The control offered by freehold ownership is often a significant motivation for purchase. Ownership avoids the requirement to consult the landlord and comply with lease obligations, in particular the need to obtain consent to alterations, alienation and change of use. This is often a strong influence for core business properties where flexibility in occupation can be of central importance. The management burden of landlord/tenant liaison and the professional fees related to consents and negotiations are also a disadvantage of leasing.

  Ownership can make disposal easier as there are a greater number of potential disposal routes, including sale and sale and leaseback as well as various leasing options, and the need to obtain landlord's approval is eliminated.

  Freehold ownership does not confer unrestricted occupational rights. In addition to any restrictive covenants in the title, statutory regulation, in particular planning regulations, can limit your use and occupation of the property.

3.1.2 Finance lease

The finance lease combines aspects of leasing and purchase and is in effect a means of financing a purchase. The finance lease is a long-term loan which works much like a capital repayment mortgage. The property is owned by the lender but the lessee retains the risks and rewards of ownership and the lease payments are more akin to debt servicing than to rent. They are calculated on the basis of the tenant's covenant and do not normally reflect the investment value of the property. The tenant often has a buy-back option at the original capital price at specific points during the term. The structure is often used in sale and leaseback transactions.

Until 1984 one of the main motivations for occupiers adopting this structure was that this type of debt was not shown on the balance sheet. This changed with the publication of *SSAP 21 'Accounting for leases and hire purchase contracts'*, which requires finance leases to be capitalized in the occupier's accounts (Section 1.5). Now the main rationale of entering into a finance lease is that the finance lessor may be able to benefit from capital allowances and enterprise zone allowances, savings which will normally be factored into the rental payments (Section 1.6). Other advantages of this structure to an occupier may include the following.

- The rental payments can be fixed at commencement of the lease or can fluctuate. This may accord with business strategy: fixing rental levels mitigates risk and relating occupancy cost to the cost of borrowing can help in business planning, notably where revenues are related to interest rates.
- Where the occupier has a purchase option they have the potential to benefit from capital gains. This benefit may be reflected in higher rental payments.
- The occupier benefits from most of the control and flexibility associated with freehold ownership, as the finance lessor's principal interest is in the covenanted stream of income and the lessee will be exercising the purchase option, so the lessor does not need to impose the restrictions normally required under a lease to protect investment value.

3.1.3 Licence

The simplest basis for occupation of property is a licence. This is a permission to occupy or use land or property and confers very few rights on the occupier. It is a personal right and cannot usually be transferred to another occupier. The occupier enjoys no security of tenure. It is usually used for temporary or short-term occupation or where occupation is shared with another party. The advantages of this form of tenure is its flexibility and the speed with which the documentation can usually be completed.

An agreement will not be interpreted as a licence simply because is labelled as such. A tenancy may be construed if the occupier:

- enjoys exclusive possession; and
- is given an express or implied right to occupy for a defined term; and
- is obliged to make periodic payments.

If you are unsure whether the arrangement is a tenancy or a licence it is important that you take legal advice so that you are able to protect any rights you may have.

3.1.4 Tenancy

The main types of tenancy are as follows.

(a) Tenancy at will

This is the most flexible form of tenancy. Under this arrangement the owner grants consent to the occupation and may specify obligations such as the payment of rent, restriction on use and obligations to repair. The tenancy can be terminated by either party at any time and there is no security of tenure. This is often the tenancy which is used (or may be implied) where occupation is granted pending the agreement of lease terms.

(b) Periodic tenancy

A periodic tenancy continues from period to period until it is terminated by a notice to quit by either party to take effect at the end of the period. The period can be chosen but is commonly a month or a year. If the period is not expressly defined it will be construed as the period by which the rent is expressed or calculated (not by which it is paid, if this is different). Periodic tenancies enjoy security of tenure under the Landlord and Tenant Act 1954, but the tenant cannot request a new lease and must wait for the landlord to terminate the lease (Section 5.2).

(c) Lease

This is a fixed term tenancy for a period of any length. Leases are often described in shorthand by reference to the obligations of the tenant as follows.

- Full Repairing and Insuring (FRI) lease. The tenant occupies the whole of the building and is responsible for all repairs and insurance. The landlord does not usually have any involvement in running the building. Sometimes the parties agree that the landlord places the insurance cover and the tenant reimburses the premium.
- Internal Repairing and Insuring (IRI) lease. The landlord is responsible for maintenance, repairs and insurance of the structure and exterior of the building and the tenant only has obligations in respect of the internal demised area. This is commonly used where the lease is for a short period and it would be inequitable for the tenant to pay for large items of expenditure.
- Effective FRI lease (Internal Repairing and Insuring lease with a service charge). This is the usual structure in a multi-occupied building where the landlord manages the common areas and maintains, repairs and insures the structure and exterior. Tenants contribute towards these common costs in proportion to the size of their occupation by way of a service charge.

Business leases granted for a term greater than six months are protected by the security of tenure provisions of the Landlord and Tenant Act 1954 unless they are specifically excluded by agreement and a court order. On expiry of the term or at a date on which the landlord can break the lease, the Act allows the tenant to hold over and apply for a new lease. The landlord can only regain possession of the premises on restricted grounds (Section 5.2).

3.2 GENERAL GUIDANCE ON NEGOTIATING TERMS

There are a number of general points to always keep in mind when negotiating the transaction and legal documentation. These are commonsense points and will probably not be new to you. Keeping them in mind at all times during the negotiations will help you to negotiate effective documents.

3.2.1 Assume the worst when negotiating

Although you will have evaluated the landlord and will hopefully be satisfied with his approach, for the purposes of drafting you should always assume that the relationship will deteriorate, or that the landlord will change, and that at some stage you will have to rely on the protection afforded by the document. By working with the expectation that you will have to rely on the drafting, you ensure that it is as strong as possible to protect you in the worst eventuality.

3.2.2 Beware of personal rights

Landlords will sometimes concede a point on the condition that it only applies for the original tenant and not for any assignee. At the time of the transaction assignments may be an unlikely prospect and the last thing in your contemplation. Business plans can change quickly, however, and the tenant should avoid personal rights if they would prejudice the marketability of the lease. If a personal right is to be accepted, it is important to agree that the tenant's group companies will benefit from the right if there is an intra-group assignment. You should ensure that any personal rights are not personal as far as the landlord is concerned and will bind his successors.

3.2.3 Think to the long term

You should not focus exclusively on your current needs but should draft the documentation with enough flexibility to cater for both changes in your own business requirements and the needs of any assignee or subtenant. Even where you are only taking a short lease you should aim to achieve flexibility, as the terms you agree now will affect those you can achieve at renewal, especially where it is a statutory renewal.

3.2.4 Consider the valuation implications of every clause

When considering and negotiating leases tenants often neglect the valuation implications of the drafting. The rent should really be the last term negotiated as it should reflect all the terms of the transaction. In practice, it is not possible to negotiate the rental deal separately. However, it is important that at all times the valuation implications of the terms negotiated are considered and

When negotiating think to the long term.

introduced into the negotiations. This enables the occupier to achieve the following three things.

- Any reduction in rental value is reflected in the rental deal
- By quantifying the effect on rental value of a particular term, you can avoid overpaying.
- Where the landlord is unwilling to agree a term which benefits you, an offer to increase the rent paid may facilitate agreement.

In addition to the original rental deal, the terms of the lease will have an impact on the rent at review, if there is one, and potentially at renewal. If there is a deviation from standard market practice it is important to determine the impact on value and the potential long term cost in order to decide whether the clause is beneficial overall. The professional team will alert you to the valuation impact of a clause where this is significant, but it is wise to be aware generally of the issue and ensure it is considered in all cases.

With these four principles in mind, you can move on to the specifics of the transaction and the documentation. While you will be advised on these issues by your professionals, an understanding of the issues and the types of clauses that may be negotiable will enable you to take a proactive role in the process and make effective informed decisions.

3.3 PURCHASE TERMS

3.3.1 Price

The purchase price will be the main subject of the negotiation and will depend on the market value of the interest. Capital valuation is discussed in Section 1.4.2. The structure of the payment can vary. In most cases where you are purchasing the property for your own occupation a simple capital payment on purchase will be agreed, but other structures can be adopted, such as an initial payment of a capital sum with subsequent payments relating to income or development profit.

3.3.2 Title issues

Investigating title is necessary to ensure that the vendor has good legal title to pass to you and that it is satisfactory for your business requirements. The site might be subject to existing covenants in favour of other land restricting the use or development of the site. A common example is a covenant against the sale of alcohol. Where the site comprises only a part of a larger site, perhaps a business or retail park, the owner may impose new covenants.

Even if the **restrictive covenant** does not cause a problem for you, bear in mind the possible requirements of future occupiers if you decide to dispose of the property. Not all covenants are enforceable but your lawyers can advise on whether the particular covenant can be enforced. If there is some doubt about enforceability it may be possible to insure against the risk with a defective title indemnity policy. It may well be appropriate to seek payment of the one-off premium from the vendor. It may be possible have a restrictive covenant removed or varied by a procedure at the Lands Tribunal. Alternatively, a formal release can be sought from the person with the benefit of the covenant.

The site may require use of third party land for access or essential services and these need to be procured on satisfactory terms. For example, a drainage right might be personal to the vendor or include onerous maintenance or payment obligations. Remember to consider both the period of your occupation and that of any future occupier.

3.3.3 Environmental issues

Environmental legislation can require owners and occupiers to clean up contaminated sites in certain circumstances. The costs of clean up can be substantial. Where you are purchasing a site and a risk of contamination is identified, the cost, valuation and occupational implications mean that this will be an important issue to be addressed in the transaction. The greatest risk of contamination applies in the case of industrial land and property, but there can be risks associated with other types of property, especially where there is a possibility that the site was previously used for an industrial process.

The options are, broadly:

- to accept the risk of the potential occupational problems and cost exposure associated with remedial works; or
- to obtain an indemnity from the vendor for costs and liabilities arising from the contamination; or
- to reflect the valuation and occupational issues through an adjustment to the price; or
- to arrange for remedial works to be carried out prior to, or immediately after completion.

The first option is not normally to be recommended. In some cases speculative investors or developers will accommodate contamination risks without explicitly reflecting them in the transaction, but for the occupier this can be a dangerous course of action. Contamination can have a significant impact on capital value; the impact can be large even if contamination is not confirmed but is only suspected. It is essential that you obtain professional advice on the matter and are clear on the valuation implications as a basis for price negotiations.

The ideal resolution is usually an obligation on the vendor to clean up the site prior to occupation together with a reduction in the purchase price to reflect the residual risk, but this is not always practical and may not be negotiable. You will need to find a solution which is acceptable to the vendor and which eliminates, or at least limits, your risk. If it is not possible to agree terms which satisfactorily mitigate the risk, you will need to consider whether to proceed with the transaction.

The main issues to be addressed in the negotiations are:

- whether remedial work is justified and, if so, the timing of the work, the standard to which it is to be carried out and responsibility for the cost;
- indemnities in relation to costs and liabilities incurred by the purchaser arising from the contamination;
- warranties relating to the vendors compliance with environmental and health and safety laws and the absence of problems arising from any contamination.

While standard public liability and property risks insurance policies normally exclude most pollution risks, it is usually possible to obtain separate insurance cover for some contamination risks. This will not normally obviate the need to address the matter in the transaction as the policies only cover certain risks and will not normally be available for sites with a high contamination risk. It may be appropriate to seek a reduction in price to reflect the additional insurance cost.

3.4 LEASE TERMS

The professional team should always be consulted and fully involved in the process and the information and advice provided here should only be taken as a starting point.

The terms of the lease are negotiated between the landlord and tenant during the Heads of Terms negotiations and legal drafting. Until the late 1980s it was common for lettings to be on the basis of the **Institutional Lease**. This is the term that is used to describe the lease structure based around the requirements of institutional investors who dominate the property investment market. The Institutional Lease includes certain structural elements, such as a term of approximately 25 years, upwards only rent reviews at five-yearly intervals, and no options to contract, expand or break. The drafting of other clauses caters for the concerns of the institutional investor by minimizing their outgoings and risk.

The Institutional Lease is often not ideally suited to the needs of occupiers as a result of its inflexibility, and in the weaker property market of recent years shorter terms and break options have become more common. Many of the elements of the Institutional Lease, such as five-yearly upwards only rent reviews, still prevail however, despite the reversal in the negotiating strength of the parties.

The occupier should be wary of agreeing terms simply because they are commonly included in standard form leases and should consider every clause in detail to assess its business implications. You will not always be able to agree the terms you want; this will depend on the relative negotiating strength of the parties. However, by being aware of the possibilities you will avoid some of the unsatisfactory terms that are commonly taken for granted. The main issues to be considered are illustrated in Figure 3.2.

A summary of the issues and advice on good practice is contained in *Commercial Property Leases: Code of Practice* (see Further Reading). This provides detailed information and helpful advice on lease terms and its recommendations should be adopted wherever possible.

3.4.1 Length of term and options to break, expand and contract

The choice of length of term and requirements for options to break, expand and contract depend on business requirements and the need for flexibility in the

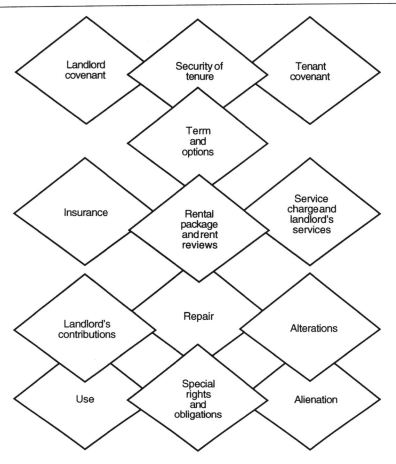

Figure 3.2 Lease structure and terms.

short and long term. In general, landlords prefer the term to be as long as possible without any option for the tenant to break or contract, as this improves the security of income. In a buoyant market the landlord may insist on a landlord operated break option. Tenants' options to expand and contract affect the investment value of the property as they restrict the landlord's ability to let space. They are therefore generally resisted by landlords.

The ability to negotiate a short term or options is dependent on the demand/supply relationship for the property and the negotiating position of the parties. In the weak property market of recent years tenants have been able to negotiate relatively flexible leases, with five- to ten-year terms becoming reasonably common. Landlords are usually reluctant to agree terms of less than five years although this depends on the demand for the property.

Longer fixed terms have some advantages, such as the depreciation of any fit-out works over a longer period and the savings in time and money from avoiding a relocation. Many tenants, however, prefer the flexibility that short terms and break clauses offer. If you want flexibility there will be a price to pay, whether directly on the financial terms or indirectly in other terms which may have to be conceded in return for options to expand, contract and break. It is essential that in the planning stage the requirements for flexibility are rigorously analysed to ensure that flexibility is of sufficient value to justify its cost. When attempting to negotiate expansion, contraction and break options the cost to the landlord of the option should be estimated as the landlord may attempt to recoup this cost in some form in the terms negotiated. This analysis comprises valuing the landlord's interest with and without the option and your professional advisers will be able to assist in this work.

You should remember that leases are usually, although not always, disposable, so you are not necessarily tied into the full obligations of the lease for the full contracted term. If your business needs change you will normally be able to assign or sublet the space, although this may be costly, especially if there is a change in market conditions, and you may be left with a continuing risk of further liability. The more you reduce the landlord's risk, the lower the price you will pay now, in theory. Flexibility can be costly – make sure you are not paying for something you do not need.

Tenant break options are now becoming relatively common. They can take the form of one-off options or rolling options. A rolling option provides a period during which the tenant can elect to break. It is favourable for tenants as it allows you a longer period in which to decide whether to break and provides a high degree of flexibility. Landlords are reluctant to agree break options, particularly rolling options, but have recently been forced to do so to achieve lettings. A break clause has a similar effect on investment value as a lease expiry but can be slightly preferable to the landlord as there is a greater likelihood of a tenant vacating at expiry than at a break option. A break option can also be preferable to the tenant as it gives you the option to remain on agreed terms, which you may not have with an expiry.

If the lease is to contain a provision for rent review, it is sometimes advisable to negotiate a break clause even if this is not required for business reasons, as an escape route in case the rent is increased at review to an unacceptably high level.

Options to expand and contract are relatively unusual but can be beneficial to a tenant if long-term business plans are uncertain. A right to contract takes the form of a surrender right on a floor or self-contained area. The area needs to be capable of separate occupation and the landlord will resist the right where the area is small or of an unusual shape as this will prejudice marketability. Expansion rights can be important where contiguous occupation will be required in the event of business growth but they can be difficult to negotiate as they can severely prejudice the marketability of the space affected.

Pre-emption rights or **rights of first refusal** can be easier to achieve. These require the landlord to offer space to the tenant in the event of another party making an offer. The tenant is usually entitled to the space on the terms offered by the other party. The timing of the option cannot be controlled and may not fit in with business plans. It is important that these clauses are drafted carefully to ensure they are effective. In particular, the landlord should have an obligation to re-offer the space to you if the principal terms of the transaction with the third party change. This prevents the landlord circumventing the clause by offering the space to you on unfavourable terms and then agreeing different terms with a third party.

Landlords can in some cases be unwilling to agree short terms, break clauses and other options because of their detrimental impact on investment value. Where flexibility is a key requirement and the landlord is unwilling to accept a flexible lease it is sometimes helpful to analyse the issue from the landlord's perspective to try to establish the impact the options will have on the value of the landlord's interest. You may then be able to offer a higher rent or other concessions to compensate them for the damaging effect of the particular option required.

When agreeing options in the Heads of Terms negotiations, it is important to agree the precise terms of the clause and not to leave this to the drafting stage. Many standard **break option clauses** restrict the operation of options by making them **conditional** on compliance with your obligations under the lease. Where this type of clause applies, you can forfeit the right to break where you have not strictly complied with all the terms. This is potentially very dangerous as some obligations can be difficult to perform strictly, for example repairing covenants can be difficult to assess until after service of the landlord's schedule of dilapidations. Potentially, you could lose your right to break simply because an accounts clerk pays the rent a few days late because of an oversight.

The clause can cause significant practical difficulties as a result of the potential uncertainty over the right to break. It can prevent the tenant from proceeding with the acquisition of alternative premises and from challenging the landlord if a breach of covenant is asserted. In order to be safe, you may have to go beyond your obligations under the lease. You will not have the option of dealing with dilapidations and reinstatement obligations by way of a financial settlement after expiry, which is the way they are commonly dealt with.

If it is not possible to avoid the restriction altogether, a compromise solution is to ensure that only 'material' breaches are covered together with a requirement that the landlord notify the tenant of any breaches a reasonable time (say six months) prior to date the break notice needs to be served. This may still leave you exposed because the break notice may be invalid if a breach arises after it is served. The consequence could be losing the break right having already committed on a new property.

Another important drafting issue that should be addressed at the Heads of Terms stage is the **period of notice** required by the tenant for exercising an option. Notice periods are commonly between six months and a year and the landlord will usually try to agree the longest possible period on a tenant's break option to assist him in the marketing of the space. Where the lease is not contracted out of the security of tenure and compensation provisions of the Landlord and Tenant Act 1954, the procedures set out in the Act must be followed for a landlord's break option, but the tenant will normally only need to comply with the contractual notice provisions (Section 5.2.7). Tenants should consider the implications of agreeing long notice periods as it may be impractical to commit to vacating until the acquisition of alternative premises is completed.

3.4.2 Security of tenure

Most business tenancies confer security of tenure under the Landlord and Tenant Act 1954 unless they are for a period of less than six months, or are excluded from the security of tenure provisions by agreement. Where the lease is a secure tenancy the determination procedures prescribed by the Act will apply and the tenant has a right to a new lease on expiry unless the landlord can prove certain restricted grounds for possession. The landlord can only frustrate the tenant's request for a new lease if he can prove one of the permitted grounds (Section 5.2).

An agreement to exclude the protection of the Act requires authorization by a court order, which has the effect of **contracting out** the lease from the security of tenure and compensation provisions of the Act. Though it can usually be obtained routinely, it can add some time to the legal process, and there are proposals to remove the requirement of a court order. Where a lease is contracted out the tenant has no right to renew at expiry or if the landlord exercises a break clause. You should consider carefully the implications of this before agreeing to contract out:

* the cost of improvements may have to be depreciated over a shorter period;
* there may be significant disruption caused by having to vacate and undertake another acquisition, for which you will not be compensated;
* disposal of a contracted-out lease can be harder.

Where you have negotiated a tenant break option you may have to accept a contracted out lease. Without this the landlord may be in a weak position at the break date if rental values have fallen, as you may be able to use the break option to renegotiate the terms of the lease, and the landlord may be obliged to grant a new lease on market terms although it now seems unlikely that this tactic would work (see Section 5.2.7). An alternative to contracting out in this case is to agree a notice period of longer than 12 months, in which case there will be no right to renew if you exercise the option. If there is a landlord-exercised break option, the landlord will normally insist on a contracted-out lease, to ensure that he can exercise the break on any grounds and is not restricted to those permitted under the Landlord and Tenant Act.

Contracting out has valuation implications which are often overlooked by tenants. If you agree to contract out you are foregoing a right to a renewal lease or the potential to receive compensation. These rights have a value that should be reflected in the financial terms agreed. The professional team will be able to advise on value differentials between protected and unprotected leases.

3.4.3 Tenant covenant

The tenant's covenant strength translates into investment value for the landlord: the better the covenant the lower the risk and hence the yield applied when capitalizing the rent over the life of the investment. In theory, the better the covenant you can offer, the more favourable the other terms that can be negotiated, although this link has been weakened for leases granted after 1 January 1996 as a result of legislation removing privity of contract (see below). By improving the covenant offered, however, it may be possible to secure a lower rent or other financial or non-financial concessions.

You should therefore consider prior to the negotiating stage what tenant covenant you can offer. Would a group company guarantee represent a better covenant? Would another company, individual or bank be prepared to stand as guarantor? Would it be possible to offer a rent deposit?

A standard institutional guarantee clause provides for the guarantor to remain liable for the full lease term (to the extent that the tenant is liable) and sometimes for any holding over period after expiry. This includes all the covenants in the lease and not just the obligation to pay rent. It could potentially include large unexpected costs such as a large repair bill if a major item of work became necessary. The guarantor may also be obliged to take over the lease in the event of tenant insolvency.

The burden of contractual agreements, the obligations they impose on each party, cannot normally be assigned when the benefit of the contract is assigned. Under the principle of **privity of contract** the original parties to a lease continue to hold an obligation to perform their covenants after the lease is assigned either by the new landlord or the tenant. It is standard for landlords to require **direct covenants** from assignees, which means that in addition to their liability while they are the tenant, they too remain liable after they assign for the duration of the lease. The result is that landlords often have a choice of parties to pursue if the existing tenant defaults and often choose to 'cherry pick' the tenants with the deepest pockets.

The privity of contract rule applies to all leases granted prior to 1 January 1996 and to leases granted pursuant to a legal agreement made prior to 1 January, such as an agreement for lease, an agreement containing an option to take a lease, or a right of first refusal on vacant space.

All other new leases granted after 1 January 1996 are governed by the rules contained in the Landlord and Tenant (Covenants) Act 1995 which limits the application of the privity of contract principle. Specifically, the Act abolishes privity of contract for new leases so that landlords cannot pursue tenants who have assigned their leases unless they have expressly provided **an authorised guarantee agreement (AGA)**. If the lease provides that the landlord can require an AGA from the outgoing tenant there is no requirement of reasonableness (unless this is specified): the landlord can require the guarantee even if it would be unreasonable to refuse consent to the assignment. If the lease is silent on the point the landlord can only require a guarantee if it would be reasonable to refuse consent without one. The guarantee can only apply to performance of the covenants by the immediate assignee, and the landlord cannot require a guarantee which continues after the assignee assigns. As a tenant you will wish to resist an obligation to give such a guarantee or to limit the circumstance in which the guarantee can be required. You may be able to negotiate a release from the guarantee once the new tenant satisfies certain financial tests. Bear in mind however that if you are considered a good covenant the limited liability will represent a cost to the landlord that may be reflected in the other terms of the transaction, unless you are prepared to agree tough restrictions on assignment.

3.4.4 Landlord covenant

The covenant of the tenant is an issue the landlord focuses on in some detail and, as a result of its implications for investment value, governs the terms that the landlord will be prepared to negotiate. Conversely, in many transactions the tenant fails to recognize the importance of the quality of the landlord's covenant.

The ability of the landlord to perform many of his obligations is dependent on his financial strength. The landlord's ability to maintain and repair the common parts, manage facilities and undertake structural repairs will all be prejudiced if he becomes financially weak. In this situation he may have difficulty funding service charge shortfalls and obtaining funds to deal with cashflow fluctuations. The issue is of greatest significance where the landlord has an extensive role in providing services, such as on an estate where he manages all the roads and common services. However, it is also important where the landlord's services are not extensive, as failure to perform even relatively minor obligations can in some cases have significant implications for the tenant's occupation.

Where there are any doubts about the landlord's financial status it is important to include mechanisms in the documentation which will assist if the landlord is unable to meet its obligations. The main ones are:

● guarantees of the landlord's covenants by another company, usually a parent or group company. It is important that the lawyer ensures that the guarantee applies to all the landlord's covenants and remains in place if the landlord assigns his interest.

● right of set off. This enables the tenant to hold back rent (and other payments) where the landlord is in breach of his obligations. This remedy is particularly important where the landlord's financial strength is in doubt. Ideally it should be combined with a right to undertake works or run services if the landlord fails to do so, but this is usually impractical in a multi-let building. Landlords usually strongly resist set-off rights because of their damaging effect on investment value.

- escrow deposit. Monies are paid into an account by one of the parties and are held on trust until particular conditions, such as persistent breach of a repairing or other covenant, are fulfilled. In the event of the condition being satisfied the monies are released to the tenant to enable him to perform the landlord's obligation. This is most commonly used where a freehold or long leasehold interest is being purchased or in other cases where a low annual rent is reserved and therefore the set-off remedy is not practical.

You should also consider the implications of the landlord selling his interest. You will have no control over the assignee and his management style, business objectives and financial strength will affect the management of the building and your occupation. The financial problem can be addressed in two ways.

- You should try to avoid any special terms automatically ending the landlord's liability when he assigns his interest, although this can be difficult unless you are in a strong negotiating position. Your recourse to the original landlord may still be limited as a result of the provisions of the Landlord and Tenant (Covenants) Act 1995 which allow landlords of leases granted after 1 January 1996 to apply to the tenant for a release from liability on disposal of their reversion. If you refuse, the landlord can apply to the court for a release.
- You may be able to negotiate a **right of first refusal** which gives you a right to purchase the landlord's interest if the he wishes to sell. This will prevent him from selling within a given period on more favourable terms than he has offered to you. This right will only be appropriate where a purchase can be funded and would meet business objectives, which will not normally be the case where you are taking a small part of a large building. They are difficult to negotiate as they are very unattractive to landlords.

3.4.5 The rental package

The rental package is negotiated at the Heads of Terms stage along with the other main terms of the transaction. It is a common mistake to ignore the impact of the other terms on the rental value and negotiate the rent as a independent item. The rent should not finally be agreed until the remainder of the package is confirmed.

Your aim should be to find the rental structure that most effectively meets the needs of both parties. Financial analysis can be used to assist in this: undertaking analysis from the landlord's viewpoint can help to find the most mutually beneficial package and also to demonstrate to him why he should accept a particular deal. For example, it may be possible to show why he should accept a lower rent with a shorter rent-free period if this better suits your or his requirements.

In negotiating the rental package it is important to be adaptable and imaginative to develop a transaction which accommodates both the landlord's and tenant's needs. Where a landlord is cash hungry, for example, it may be possible to agree a lower rent if this is **prepaid** for a period at the start of the lease. Where the occupier has a low cost of capital this can be particularly advantageous. A **capital contribution** may be appropriate for a cash rich landlord, whereas a **rent-free periods, stepped rents** or **phased rents** are more common.

Takeover of tenants' properties is becoming common in the current market where existing lease commitments inhibit tenant relocations. The landlord takes over the leases on existing properties, leaving you free to take a new lease on his property. The cost to the landlord is reflected in the rent and other terms agreed. These arrangements offer the advantage of transferring the burden of unwanted properties to the landlord who may have the management resources to deal with them more effectively. Where takeover properties are included in the transaction, and your leases on these properties were granted before 1 January 1996, you will need to be satisfied about the landlord's covenant to be sure that the obligations will not revert to you if your landlord becomes insolvent. A right of set off is important to avoid your finding yourself having to pay rent and other outgoings on both properties.

Turnover rents can often be appropriate for retail properties. These incorporate a fixed rent at a proportion of the rental value (usually approximately 80%) with a formula for calculating an additional uplift based on the turnover achieved. In some unusual cases a turnover-only rent basis can be agreed. The base rent will normally be subject to review in the normal way and the issues discussed below under rent review apply.

The turnover structure is becoming common in shopping centres. It has the advantage to the tenant of transferring some business risk onto the landlord and you may pay a higher total rent to reflect this risk. In some cases, particularly in shopping centres you will have no choice but to accept a turnover rent as this will be the common basis for the centre. You may pay a higher total rent to compensate the landlord for the additional risk he is taking.

A similar structure to turnover rents is a **geared rent** structure which again comprises a fixed rent plus an uplift based on income. This is used where some or all of the premises will be underlet and the landlord wishes to gear the rent he receives to the rent received from the subtenant. The level of the geared rent should take into account the prospect for rental growth.

Index linked rents are unusual in the UK but common in Europe. They have the advantage of relating rental costs to general price levels and can help to match outgoings with revenues.

The most common rental structure is simple fixed rent with a rent-free period. For a ten- or fifteen-year term rent-free periods of between six months and two years are relatively common and they can be longer in particularly oversupplied markets. In some cases the rent-free period is related to the period during which the tenant will fit out and will not be able to occupy. Six-month rent-free periods for fitting out are common. When negotiating a rent-free period, be sure the period is to start from completion of the lease and not agreement of Heads of Terms; the latter can severely restrict your negotiating strength by putting a time pressure on negotiations.

In the negotiations, analysis of the **effective rent** (Section 1.4.2) should be used to evaluate different rental packages. Alternative transaction structures should be compared using effective rents to determine which is the most favourable.

It is common to apply different rental rates to different parts of the property to reflect their use and amenities. In a retail acquisition the rent is normally calculated using a zoning system (described in Section 1.4.2). For offices the value generally relates to the use, specification, extent of natural light, accessibility and proximity of facilities. The prime office rate is usually applied on the lower office floors, especially where there is no lift access; basement areas are usually valued at a quarter to half of the prime office rate.

When structuring the transaction it is important to consider potential tax implications and develop the most tax-efficient structure. Refer to Section 1.6 for a detailed discussion on these issues.

Rents quoted in marketing details are usually calculated on the basis of floor areas measured on behalf of the landlord. These floor areas are notoriously inaccurate and it is usually advisable to obtain separate **measurement** of the premises to ensure that the rent is calculated using an accurate area. The cost of the process is usually small in comparison with the rental savings which can be achieved over the term. The area should be calculated on the basis of the latest edition of the *Code of Measuring Practice* (currently fourth edition) using the definition appropriate to the type of property concerned. The *Code* details all the areas to be included and excluded and helps to minimize the disagreement between the parties (Appendix A).

There are three main ways to establish the floor area:

- Measurement by your agent or property adviser followed by negotiation with the landlord to agree the area – this can be time consuming and expensive and result can sometimes be inaccurate, based on a 'horse deal'.
- Joint measurement by landlord's and tenant's surveyors – the agents jointly inspect and measure the property. Again the result may be inaccurate as a result of the negotiation process involved.
- Independent measurement by a third-party surveyor acting jointly for the landlord and tenant – this is often the most efficient and cost-effective method as it avoids disputes and instruction of two sets of surveyors. The surveyor can be instructed by both parties to owe a duty of care to both. As such he uses his professional judgement in resolving interpretation problems to arrive at the fairest solution in accordance with the code. There can be significant time savings as the disputes that normally arise in agreeing measurements are usually avoided. The surveyor's fees can be shared between the parties but if a landlord refuses to contribute it can sometimes still be cost effective to use a joint surveyor at the tenant's cost because of the negotiation process that is eliminated.

The most cost-effective and accurate method of measurement is usually an electronic system using Computer Aided Design (CAD) technology.

It has become standard for all payments to the landlord listed in the lease, including service charge and insurance, to be reserved as 'rent'. This can enable the landlord to use all the remedies available in respect of rent in the event of the tenant defaulting on other payments, including the use of bailiffs. This is explained further in Section 4.13.1. Where the tenant is in a strong negotiating position it may be possible to exclude all non-rental payments from the **definition of rent**.

Most leases do not require the landlord to demand payment of rent although in practice landlords normally do send invoices. It is hard to negotiate such a requirement as landlords will be reluctant to limit their right to forfeit which only requires formal demand if this is expressly stated. You should pay attention to the **grace period** allowed under the lease in relation to payment of rent and other charges to ensure that it will be practical for you to comply with these. There may be a period of grace, normally between five and 21 days, before the landlord can use remedies for non-payment. If you are unable to negotiate such a provision in relation to the payment of the rent, a grace period for non-rental payments such as service charge and insurance may be sufficient. You should also pay attention to the provisions for charging **interest** on late payment; a period of grace before interest can be charged is common (usually 5–10 days).

3.4.6 Rent reviews

Five-yearly upwards only rent reviews have become common in modern leases. However, the occupier should consider carefully the implications of agreeing to these provisions. The upwards only rent review was developed to meet the needs of investors for inflation proof income. The commitment to pay rent in the future at an unknown level can have severe business implications. The rent payable will be at the market level which may not be a level that can be supported by the business. For example, rental values may be inflated if the location becomes attractive to a different type of user able to support a higher rent. Unlike in a new letting situation, as the tenant you will not be able to walk away from the negotiating table: you are contractually bound to pay the higher rent.

Even in the current market, landlords strongly resist leases without reviews at least every five years. However, rents fixed for as long as 15 years have been seen on some rare occasions and where the tenant's negotiating position is very strong it can be possible to avoid rent reviews. The occupier needs to decide how significant the business risk is and what the implications of a steep rental increase would be. If you consider that this is a term you wish to avoid there are ways of structuring the transaction to do so but there will usually be a price to pay. The concessions the tenant may be able to make could include prepaying the rent for the whole of the term (if the capital is available), paying a higher rent from commencement, or agreeing a fixed increase in the rent at the fifth year.

Rent reviews can be avoided by reducing the length of the term or their impact can be limited by introducing a break clause at the date of review to enable you to negotiate the rent freely in the open market. Where a break option is negotiated it is important to ensure that the notice provisions for the rent review tie in with the timing for service of the break notice in order to ensure that you have a clear idea of the market rent before deciding whether to exercise the break. It is advisable to provide that the landlord's notice must be served at least three months before the deadline for service of the break notice and that time is of the essence (Section 4.2.1).

The standard institutional lease provides for rent reviews to be **upwards only**, and even in the current market it is hard to negotiate freely floating (upwards or downwards) reviews. The problem of upwards only reviews has been brought into sharp focus as a result of recent falls in the property market, where the review procedure has failed to protect tenants against severe over-renting. The importance of negotiating a freely floating review depends on your assessment of the likelihood of rents falling before the review date. If you expect the market to rise you may decide this is not an important issue. For long leases with several reviews or where there is a good chance of a fall in rental values it may be worth resisting the upwards only provision.

Landlords sometimes try to include a provision for a **penultimate day rent review**. This falls on the day before expiry and enables the landlord to bypass the Landlord and Tenant Act procedures for an interim rent if you hold over after expiry (Section 5.2) and the clause should normally be resisted. Depending on the wording of the review clause there may be valuation implications if the definition is materially different from the assumptions applied in determining interim rents.

(a) Valuation issues

The aim at rent review should be to reflect accurately the market rent that would be payable for the premises if they were available to let. To achieve this the rent review clause has to define the basis on which the premises would be let. This is done by imposing various assumptions including a definition of the hypothetical lease and open market rent. The following are the usual assumptions included in modern leases:

Definition of open market rent
Open market rent is usually defined as the 'best clear yearly rent at which the premises would be expected to be let in the open market', or similar wording, with the following express assumptions:

- a willing landlord and willing tenant;
- as a whole with vacant possession;
- without a fine or premium; and
- on the terms and conditions contained in the lease.

Taken together these elements of the definition aim to replicate the situation in an open market letting.

A further part of the definition is the hypothetical lease to be assumed. The rental value is dependent on the other terms of the letting and therefore for rent review assumptions have to be made as to those terms. As far as possible you should aim to ensure that the hypothetical lease mirrors the actual lease except only for the amount of rent in order to achieve a rent based on the terms which actually subsist. Some exceptions may be necessary. In particular, the length of the term to be assumed may have to be specified to avoid an artificially low or high rent as you approach expiry. From the tenant's view point it is usually fairest for the assumed length of term to be the unexpired residue as this reflects the real situation, but you may have to accept a fixed length, often ten years.

The implications of the drafting of rent review clauses can often seem remote, but they can have a significant impact on the rent paid at review and it is worth taking the time to consider them and negotiate them effectively. You should be wary of agreeing to any disregards or assumptions which will result in a rent significantly different from the figure that would be obtained if the premises were available to let. Assumptions and disregards should only be included to the extent that they are necessary to achieve a fair assessment of the market value. You should resist deviations from the actual lease as far as possible, such as an assumed user clause significantly wider than the actual one in the lease or alienation rights from which you will not in fact benefit.

The following assumptions and disregards are also commonly included in rent review clauses.

Assumptions
The modern lease commonly includes the following assumptions.

- The premises are fit for and fitted out for immediate occupation. This is designed to ensure that concessions granted to tenants in a market letting to reflect the fitting-out period are ignored. The tenant should be concerned to ensure that there is no clause included which requires other rental concessions in the open market to be ignored. Rent-free periods and other incentives which do not relate to fitting out are part of market transactions and should be reflected in the analysis of the comparable evidence used to establish the rental value of the demised premises. To ignore these would distort the level of rent applied to the premises.

  Recent court decisions have demonstrated a prevailing concern to avoid artificial assumptions and a general inclination, where possible, to uphold an interpretation of the lease allowing concessions to be reflected. The occupier should resist any clause which attempts to move away from a realistic assumption on incentives.

- The premises are in a good state of repair and decorative condition. This is applied where the tenant has the repairing obligation and it is designed to avoid the landlord being penalized in the rent for the tenant's breach of a repairing covenant.
- All the covenants on the part of the landlord and tenant have been fully performed. This is a common assumption but you should consider the implications. If the landlord has not fulfilled his obligations and for example the common parts are in a poor state of repair, this will not be reflected in the rent and you will be paying for something you are not getting. You would normally be advised to seek to remove the reference to the landlord's covenants to avoid this problem. The justification for the clause is that the rent fixed at review will apply for the full review period (normally five years) and the landlord should not be penalized for several years for a breach that may be remedied immediately after the review.

Disregards

It is standard to include a provision requiring that the following items are to be disregarded in establishing the rental value at review.

- any effect on rent of the tenant's occupation;
- any goodwill attached to the premises by virtue of the occupier's business; and
- any increase in rental value resulting from tenant's improvements carried out other than in pursuance of an obligation to the landlord. This prevents the tenant paying rent in respect of works it has undertaken at its own cost, but works that you are obliged to carry out, such as any fire precaution works or other works that you contracted to undertake, will be reflected at the review. If you envisage undertaking significant works of this kind it may be advisable to resist the exclusion.

(b) Determining the open market rent – procedures and notices

The procedures for initiating and agreeing the rent review are usually set out in the lease. Older leases tended to include a detailed **notice mechanism** including a provision for the landlord to notify the tenant of the proposed rent and the tenant to serve a counter notice. The landlord could lose the right to review the rent if he failed to serve the notice correctly just as the tenant could lose the right to dispute the landlord's assessment of rental value.

However, unless time is of the essence there is no penalty for late service of notices and the landlord does not lose the right to review the rent. The time limits in notice provisions are only interpreted strictly where the lease specifies expressly that time is of the essence, or where this is implied, for example by the existence of a break option at the same date as the review. This is explained in further detail in Section 4.2.1.

In modern leases it is more common to provide simply for the rent to be agreed between the parties without any specific notice provisions. There is usually a time limit for negotiations, after which the matter can be referred to a third party. In general, occupiers should aim to ensure that the rent review procedures are detailed enough to minimize the uncertainty that can result if the review is not settled swiftly, without establishing unnecessarily rigid procedures.

Where the timing of the rent review is close to the date of a **break option** you will need to ensure the review procedure will provide you with enough information at the time of the break notice in order to make an informed decision on exercising the break option. One solution is to provide for a moving date for the break notice linked to settlement of the review. This ensures that you are fully informed when you have to serve the notice but is often resisted by landlords because of the uncertainty created. A compromise can be to require the landlord to have served a notice proposing a rent by the date of the break notice.

Where the rent cannot be agreed between the landlord and tenant, the lease will normally contain a provision for referral of the review to a **third party**. The lease may contain specific procedures, including time limits for submissions from the parties and payment of the third party's fees.

The occupier should consider whether the lease should provide for the third party to act as **expert** or **arbitrator**. This issue is explained in more detail in Section 4.2 but in summary an arbitrator is obliged to decide between the submissions of the parties but an expert determines the rent according to his own professional view and may not base it on the evidence or arguments of the parties unless there is a specific requirement in the lease. Broadly, the implications of this are that in a rising market the rent may be determined at a higher level by an expert, who is not governed by historic evidence, than by an arbitrator; in a falling market the opposite can apply. Sometimes, the clause gives the landlord the choice of expert or arbitrator at the time of the review. Occupiers should strongly resist this type of clause and an assessment of the best option should be made at the time of the acquisition based on professional advice.

3.4.7 Construction issues

The construction issues involved in an acquisition can range from very minor items in a small acquisition in an office building to a complex range of issues for a pre-let of uncompleted space. In general, for both large and small acquisitions, the main issues to be considered are the landlord's contributions to remedy base building inadequacies, the repairing obligations of the landlord and tenant, and tenant's fitting-out works.

(a) Landlord's contributions to base building

A building can be let finished to one of three types of specification, **shell and core**, **standard developer's finish** or **fitted out**. Table 3.1 lists the items included within the three definitions. The standard developer's finish is commonly referred to as a **Category A finish** while the fitted out specification is known as a **Category B finish**. Table 3.1 provides shows broadly the items usually included in the different definitions, but the terms are not always used in the same way, and will be adapted to the building concerned; in each case it is important to clarify exactly what is meant. The items listed in Table 3.1 under the definition of developer's finish will all apply for prime offices, but some items will be excluded for other types of space where the market norm dictates a reduced specification. The Category B fit-out works can vary widely depending on the tenant's requirements.

As a general rule, new office and retail premises are usually let on a Category A basis. Shell and core is commonly used for letting new industrial premises

Table 3.1 Construction definitions

Definition	*Basic construction items*	*Category A items*	*Category B items*
Shell and core	Structure and core ductwork for services only – floors and ceilings left bare at the slab	None	None
Standard developer's finish	As above	Raised floors, suspended ceilings, fire protection systems, trunking, electrak and floor boxes, carpets, basic decorative wall finishes, window blinds, services (extending from the core ductwork) including lighting, heating, air conditioning, sanitary appliances etc.	None
Fitted out	As above	As above	Fixtures and fittings, tenant's plant and equipment, partitioning, additional decorative finishes, alterations to services and additional tenant's facilities such as kitchens and WCs

and B1 premises (business use, i.e. mixed office and light industrial), and is sometimes used for retail and office buildings. Second-hand buildings are often let fitted out but in many cases the fit out has to be removed and the premises refitted.

It is common for lettings of office and retail premises that some items in the Category A specification are not undertaken by the developer prior to letting and are undertaken subsequently by the tenant and paid for by the landlord. These are items which are suited to completion as part of the tenant's fit out, such as installation of floor boxes and trunking, carpets, some decorative finishes and window blinds. Additionally, where enhancements to the Category A items are required because they fall below the standard specification it is usual for the landlord to contribute to these. Common examples are replacement of carpets, and alterations to the ceiling, raised floors and lighting.

The **Category A contributions** reflect the fact that the Category A specification is the standard basis for lettings and is reflected in the valuation for determination of the rent at both the letting stage and subsequently at review. To ensure that the rent fairly reflects the value of the premises the landlord should pay for those items where the property falls below the common standard.

In the site selection stage the property evaluation should identify the Category A shortcomings for which the landlord should pay. As part of the counter proposal in the negotiation stage, the occupier can claim from the landlord an amount for each of these items. This may be a total figure or an amount per square metre or per item.

In some cases you may also be able to secure landlord contributions to Category B items. It is now becoming more common for landlords to provide a fully fitted-out, or 'turnkey' specification which includes all tenant's fixtures and fittings. This can have the advantage of transferring the responsibility for all construction works to the landlord but equally can reduce your control over the end product and the cost.

All relevant documentation needs to be obtained from the landlord by the construction professionals to ensure that any deviations from the design specification are identified. In particular, newly constructed buildings should normally have a **base building specification** which defines the design specifications and performance criteria of the demised premises and common parts. This will include a description of the structure, finishes and services as well as detailed design criteria for services.

As part of the technical evaluation the project manager, building surveyor and mechanical and electrical engineer should check the base building specification against the building and identify any inaccuracies or shortfalls. Where these are material they should be addressed as part of the transaction negotiation. The base building specification can then form part of the legal documentation as a record of the state of the premises at the time of the letting. This has significant long-term benefits, particularly at rent review when it can be used to clarify the specification to be valued, and on exit when it confirms the finishes to be reinstated. Inclusion of the base building specification in the documentation can save considerable time and money by avoiding uncertainty and disputes in the future.

Where the tenant's fit-out works will interfere with the services in the building, such as the air conditioning and electrics, it is particularly important to ensure that the services perform in accordance with the base building specification so that you will not be held responsible for any problems later. The fit-out contractor will normally wish to satisfy himself on this as he may be liable if the performance criteria are not achieved after fit out. He may need to undertake a specialist survey and if so they should liaise with the surveyor or mechanical and electrical engineer to coordinate the surveys and minimize cost.

(b) Repair

Modern commercial leases oblige the tenant to maintain and repair the demised premises. The repairing covenant varies but usually includes the obligation to put and keep in good and substantial repair or good tenantable repair. Deviations from this wording should be discussed with professional advisers to ensure that they do not give rise to an unduly onerous obligation. In particular, where the premises are in a multi-occupied building you should ensure that there is no obligation to undertake any structural repairs.

The premises will normally have to be kept in a condition that would be acceptable to the average tenant taking the premises, having regard to the age, character and locality of the premises at the date of the lease. Refer to Section 4.5.1 which discusses the meaning of repair in more detail.

It is important to note that the covenant to keep the premises in repair includes an obligation also to put them into repair even if this is not stated. The tenant will be obliged to undertake works which put the premises into a better condition than when let, if they were in disrepair at that stage.

It is often advisable to limit the repairing obligation, particularly in the case of second-hand properties and short leases and there are a number of ways this can be done.

- Exclusion of damage by **insured risks** (except where the insurance is vitiated by the tenant). This only applies where the landlord insures the demised premises but where this is the case it is a common exclusion and should be insisted upon.
- Limiting the tenant's liability to keeping the premises in the state that existed at the time of the letting. A **schedule of condition** (normally a written schedule with photographs) is needed to record the state of repair of the premises at the time of the letting. This is a commonly used method, particularly for old second-hand buildings.

- Exclusion of **patent defects** (faults in the design, workmanship or materials used in the original construction of the building, which is identified by a surveyor in the building survey). The specific defects can then be itemised at the survey stage in a **defects schedule** and documented as exclusions from the repairing covenant.
- Exclusion of **latent, or inherent, defects** (faults in the design, workmanship or materials used in the original construction of the building, which were not apparent at the time of the letting). Landlords strongly resist this exclusion because of the open-ended nature of the cost, but this is precisely the reason why you should insist on it.
- Exclusion of **fair wear and tear** (damage caused by the tenant's ordinary and reasonable use of the demised premises). This is a relatively uncommon exclusion but is sometimes relevant for short lettings, particularly of second-hand fitted-out space, or if the building is likely to be demolished at the end of the term.

In multi-occupied buildings, an exclusion of liability for patent and/or latent defects should be accompanied by an exclusion of the tenant's service charge liability for repair of defects undertaken by the landlord. It is also important that the landlord has an obligation to rectify any defects as these can have an impact on the use and occupation of the premises. The documentation should provide procedures for rectification by the landlord on notification of a defect by the tenant. Ideally there should be a right for the tenant to undertake the work and recover the cost if the landlord does not comply within a reasonable timescale, but this is often not practical in a multi-occupied building.

In some cases patent defects will need to be rectified by the landlord prior to occupation to avoid disruption to occupation or fit out. In this situation strict time limits may be required to prevent a delay in occupation. In some cases, where delays in fitting out and taking occupation could have serious business consequences, it may be appropriate to incorporate a procedure into the documentation which enables the tenant to rectify new defects arising in the pre-fit out or fit out stage. The legal drafting necessary to make these types of clauses effective is often complex and the tenant should take the advice of an experienced construction or property lawyer.

Landlords often strongly resist any limitation on the tenant's repairing obligation; it can have a significant impact on the investment value of their interest because of the potential cost exposure. However, it is often important to elevate this as a central issue in the Heads of Terms negotiations, as the costs to the occupier of an unlimited obligation are potentially substantial. In particular, few tenants recognize the risk to which they are exposed in relation to patent and latent defects. The costs of rectifying these defects can be disproportionate and the exclusion is usually worth fighting for.

In a multi-occupied building where a **service charge cap** can be negotiated, this can be an alternative means of limiting the tenant's exposure for structural and other major repairs and this is sometimes easier to agree than a defects exclusion. Other ways of limiting the tenant's service charge liability are discussed in Section 3.4.8 below. Additionally, where the building has been recently constructed the landlord may be able to procure **collateral warranties** from the design and construction team, although the protection these offer can be limited as a result of limitations on liability. **Inherent defects insurance** may be available in some cases (Section 4.4.1).

(c) Alterations

In most cases you will wish to fit out the premises to meet your occupational needs. These are the Category B works. Even fitted-out premises will often require some alterations. On acquisition the tenant needs to consider both the alterations covenant in the lease and the terms of any licence to alter granted by the landlord at the time of the letting.

Alterations covenant

In broad terms, there is no common law restriction on a tenant's right to make alterations, but it is standard to include a covenant prohibiting alterations without landlord's consent. The clause normally allows certain alterations with the landlord's consent subject to the proviso that consent is not to be unreasonably withheld. Where this proviso is not included the Landlord and Tenant Act 1927 imposes restrictions on the landlord's right to refuse consent to improvements. This is explained in further detail in Section 4.6.

When it comes to negotiating the alterations clause, you should ensure firstly that any prohibited alterations are limited to items which will affect the structural integrity of the building, usually specified as principal or load-bearing walls, floors, columns etc. Do not agree, if you can help it, to a general prohibition of any structural works as this will be interpreted widely to cover any item comprising part of or attached to the structure. Additionally, you should attempt to secure the following protections.

- In some cases the landlord will agree to the alterations clause allowing some minor non-structural works without consent, often restricted to demountable partitioning,. This can save significant time and cost by eliminating the need to obtain the landlord's consent for every minor item of work.
- It is helpful to insert a clause requiring that the landlord's consent is not to be unreasonably delayed. A specific time limit for responses is yet more favourable but will normally be resisted.
- It is advisable to limit the landlord's right to reimbursement of costs in processing applications for consent to 'reasonable and proper' costs only. Without this restriction, the tenant has little scope for challenging the level of costs charged.

The procedures for obtaining landlord's consent are discussed in Section 4.6.2.

Licence to alter

Where you will be undertaking substantial fit-out works you should arrange for detailed plans to be prepared as early as possible in the acquisition process (normally at the negotiations stage) to ensure that consent to the alterations does not delay the acquisition. The landlord will need time to review the drawings and take advice from consultants where necessary. They may then propose design changes which will need to be discussed between consultants and incorporated into the drawings. Once the design is approved the consent can be documented by a letter licence or formal licence to alter. Sometimes the licence to alter is completed in draft and annexed to an agreement for lease to be completed retrospectively once the alterations are completed.

The tenant is in a stronger position at the acquisition stage to dictate the terms of the licence to alter than when a licence is granted during the term. During the lease the landlord can require that all works which do not add to the letting value of the premises are reinstated at the end of the lease. **Reinstatement** of the fit-out works should be addressed as an issue at the Heads of Terms stage. Where the fit out includes base building (Category A) works there is a strong argument for excluding these from the tenant's reinstatement obligation. These are items such as floor boxes, trunking, carpets and window blinds and some enhancements to base building items such as lighting, raised floors and ceilings. Additionally, the tenant should argue to exclude any Category B 'improvements'. These are items which increase the rental value of the premises and are only those which would represent a benefit to the average tenant. They can include such items as enhancements to air conditioning, ceilings and lighting, and sometimes installation of other facilities such as WCs and kitchens.

It is often difficult to persuade a landlord to accept any limitations on the reinstatement obligation, but it is usually worth trying to achieve as many exclusions as possible as the reinstatement costs saved can often be high. Where you do have to agree to reinstatement, you should aim to make this conditional on the landlord to informing you of the items he requires reinstating prior to expiry.

It is worth noting that where the fit out works include improvements it may be advisable to make an application to the landlord for approval under the Landlord and Tenant Act 1927 in order to be entitled to **compensation** at the end of the term (Section 4.6.3).

3.4.8 Service charge

In the case of a multi-occupied building, the management of the common parts, facilities and services as well as repair and maintenance of the structure, exterior and common areas is normally undertaken by the landlord. The cost of these services is usually shared between the tenants in the form of service charge.

There is very little statutory regulation of commercial service charges and the tenant is therefore reliant on the drafting of the service charge clause. This defines the landlord's service charge obligations and the costs recoverable from the tenant.

(a) Cost recovery

There will normally be a list of items, often appearing in a schedule to the lease, for which the landlord can recover costs. It is normal for these to include specifically all costs expected to be incurred together with more general wording giving the landlord discretion in determining what services to provide and charge for. The services list in a modern lease will normally include the following items.

- Repairs and maintenance of the structure, exterior and common areas;
- Air conditioning or heating;
- Heating, lighting, cleaning of common parts and window cleaning;
- Provision of supplies and hot and cold water;
- Disposal of refuse;
- Landscaping;
- Maintenance and, in some cases, renewal of plant and machinery, such as lifts and air conditioning or heating equipment;
- Reception, security and management staff, staff accommodation and uniforms;
- Costs of running management accommodation, including business rates and telephones;
- Enforcement of covenants;
- Preparation and audit of accounts;
- Complying with statutory requirements;
- Insurance (usually charged as a separate item);
- Management costs or managing agents fees;
- Professional fees;
- VAT;
- Interest where borrowing is required.

In general, the landlord will make the list wider than necessary to ensure he does not suffer a shortfall. The tenant will normally have to accept this and rely on the audit process described in Section 4.3.2 to ensure that no improper costs are charged. You should however scrutinize the list for any item which could potentially be abused and possibly include a clause limiting the amount of expenditure on this item. At the least, you should obtain clarification in writing from the landlord of what is intended. Examples of this type of item are: promotion or marketing which could be used by the landlord to pass on costs from his own business activity; and renewal and replacement of plant or other which could be used by the landlord to improve the building.

Service charges – there will be a list of items for which the landlord can recover costs.

Where specific figures are inserted into the clause the tenant should ensure that these are reasonable. These types of clauses should be carefully scrutinized as they may preclude the right to challenge costs later if they fall within the specified limit. For example, it is common to find an upper limit on management fees expressed as a percentage of the total service charge cost. This percentage should be checked to ensure it is in line with existing fee bases, and you should aim to achieve a specific exclusion of landlord's routine rent collection, portfolio management and letting costs.

The landlord will wish to include 'sweeping up' clauses to ensure that unanticipated expenditure is recoverable. The tenant should ensure that these are not drafted too widely. It is important that the lease provides that the landlord can only recover expenditure reasonably and properly incurred. You may also be able to negotiate specific exclusions at the Heads of Terms stage, as follows.

- Improvements. The landlord will often try to include the cost of improvements within the recoverable service charge expenditure. These are wholly new items which would improve the amenity of the building. You should normally resist such an inclusion: this does not preclude the landlord from undertaking these works subject to the approval of the tenants at the time.

- Patent and inherent defects. Where the exclusion of patent and inherent defects is agreed, these items should be excluded from the service charge obligation as well as the repairing obligation. The documentation needs to be carefully drafted to ensure that this exclusion is effective. Patent defects can be specifically excluded by reference to the defects schedule prepared as part of the survey. The issue is discussed further in Section 3.4.7 above.
- Replacement or renewal of major items of plant, such as boilers or lifts. It is often advisable to seek this exclusion, especially for short leases where the tenant will not benefit from the new equipment. In the absence of a general exclusion of major plant items, you may be able to negotiate the exclusion of items identified in the survey as being likely to require renewal in the short run.
- Major repairs. An alternative to a full defects exclusion is to exclude major repairs. This can be drafted by reference to an upper limit on any one item of expenditure or by reference to certain types of work. Without this exclusion the tenant is potentially exposed to very large costs and to the possibility that the landlord will undertake major works just prior to rent review, so that you pay twice for the works, firstly in the service charge and again in a higher rent.

These exclusions are often hard to negotiate and complex to draft. It may be preferable to negotiate a **service charge cap** which prevents the total cost in any one year rising above a particular level. It is sometimes only possible to agree a cap at a margin above the current cost as the landlord is often unwilling to accept an anticipated shortfall. A cap based on a five-year average can be preferable to a limit that takes effect in each year, as it takes account of uneven expenditure.

If you are taking a long lease and have any doubt about the landlord's ability or willingness to meet service charge shortfalls, you may need to consider carefully how you deal with this matter. If your exclusion may result in a significant service charge shortfall this may affect the landlord's willingness to incur costs and hence the level of service he provides. You will need to satisfy yourself that the management of the building will not be prejudiced by a limitation on your service charge contribution. If the landlord is an institution or reputable property company this is less likely to be a problem, but the current landlord may of course sell on at any time. Where you have serious concerns, you may need to negotiate, where practical, a right to undertake works to the property or to run services where the landlord fails to do so, together with a right of set off.

In rare cases the landlord may require a service charge cap to be accompanied by a service charge 'floor', or lower limit on your contribution. This should be strongly resisted as it can encourage the landlord to cut back services and make a profit.

The service charge clause will normally provide for **on account** payments to be made quarterly based on **budgeted expenditure**. The expenditure is reconciled at the **service charge year end** and **accounts** prepared confirming the total expenditure for the building. In some cases the lease requires the landlord to have the accounts audited by an independent accountant. This is a cost that some tenants resist but in general it is a protection worth paying for.

The year end reconciliation includes a calculation of the tenant's contribution based either on a specific percentage contained in the lease, or, more commonly a fair percentage calculated by the landlord. The usual modern drafting is a 'fair and reasonable proportion' and this provides flexibility. You should resist any wording which limits your scope to challenge the percentage applied. The most favourable wording, but one to which it is hard to persuade a landlord to agree, is for a fair proportion subject to an upper percentage limit.

It is advisable to include some wording defining the 'fair and reasonable' proportion. In particular, the following should normally be specified.

- The contribution relating to unlet lettable areas will be paid by the landlord and not passed on to the tenant.
- The proportion will be calculated with reference to the floor area of the premises in relation to that of the building having regard to the benefit derived from the facilities and services by the tenant and other tenants in the building. This means that if a service is used disproportionately by particular tenants they should pay in proportion to their usage not on a simple floor area basis, although it is rarely possible to ensure a perfectly fair basis where all tenants pay for exactly what they use. It is normal for tenants using the premises outside normal business hours to pay the full cost of running the services during those hours.

The tenant should ensure that the lease includes a right for the tenant to inspect invoices and accounts to check expenditure. It is also helpful to include a time limit on the landlord's preparation of accounts (usually approximately five months after the year end), but landlords often resist this. There should ideally be a time limit on charging any expenditure from when it is incurred, normally one or two years.

Most service charge clauses will include a right for the landlord to run a **sinking fund** to cover large one-off items of expenditure, but they are not often used mainly because of tax disadvantages. Some occupiers find this system unattractive because of its cashflow disadvantages and the potential to lose the benefit of the money if you assign the lease or at expiry, but they do have the advantages of enabling budgeting and avoiding large irregular payments.

(b) Landlord's services

The landlord's obligation to provide services is contained in the landlord's covenants section usually towards the end of the lease. The landlord will usually attempt to draft the lease to limit his obligation as follows.

- He may try to make his obligation conditional on the tenant paying the service charge. These clauses are often ineffective, depending on the circumstances. They are best avoided, as where they are effective they can cause problems, particularly as you may wish to withhold the service charge if it is in dispute.
- He will normally want a wide discretion on the services provided and will not wish to be liable if there is interruption to any service. Where a particular service is crucial to the tenant's occupation it may be necessary to require a more rigid obligation on the landlord. This is sometimes necessary for example where the disruption of a service will involve significant business loss, such as the provision of standby power. In this type of situation it may be essential to impose an absolute obligation on the landlord, subject only to very limited exclusions. Such obligations are generally strongly resisted.

The range and level of services required need to be considered and agreed with the landlord. Where some services are not provided or need to be extended, for example by increasing the hours of operation, this needs to be agreed and the charging basis clarified. If other tenants do not require the service the tenant may be obliged to take the full cost.

3.4.9 Insurance

Insurance has become a complex technical area and an important part of legal drafting. In order to achieve a document that addresses all the relevant insurance issues effectively, it is advisable to involve your broker or other insurance adviser at the time of the acquisition. The landlord should also be encouraged to involve his brokers and insurers. This will mean that the information you receive about the landlord's insurance policies will be accurate and that insurance issues can be addressed and resolved effectively at acquisition stage. Often insurance specialists are only brought into the process after the document is completed and there is no scope for them to assist in achieving a practical workable document.

The landlord will have significant capital invested in the property and will normally wish to retain control of the main insurance policies. In multi-occupied buildings the landlord will always place the insurance cover for the building. Where the tenant occupies the whole of the building the landlord may be prepared to allow you to insure if you wish to do so. Where you are taking a long leasehold interest in the property there is a stronger case for your placing insurance. Nevertheless, in many cases the landlord can secure insurance at competitive premiums and it is advisable to let him insure.

The landlord's policy will cover the building and the demised premises but will normally exclude the tenant's fixtures and fittings. The landlord will normally be required to insure the building for the **full reinstatement cost** for the **insured risks** which normally include all main risks, such as fire, storm, tempest, flood, earthquake, lightning, explosion and so on. You need to satisfy yourselves that the insured risks include all those you would normally wish to cover. The problem can be that the landlord may require to have full discretion in deciding which risks to include, which limits your control. You should seek a provision allowing you to add additional risks subject to availability in the market at reasonable cost.

The risks covered normally include terrorism which is currently insured under a separate government sponsored mutual pool as a result of the level of terrorist activity in recent years. In some cases cover will not be available for certain risks; for example, insurers sometimes refuse to insure for malicious damage in areas with high crime rates. The lease will normally allow certain risks to be excluded from cover either at the landlord's discretion, or where insurance is not available or economic.

Tenant's base building alterations will normally be covered by the landlord's policy but this is not always the case and the tenant should clarify the **insurance responsibilities** and agree with the landlord at the time of the letting an efficient and practical split having regard to the type of works anticipated. Although it is possible to revise the split by agreement subsequently it is advisable to establish a workable regime at the acquisition stage to avoid confusion later on.

The split should be designed to minimize complication in the event of a claim. Where the tenant's fit out includes alterations to the base building structure or services (Category A items) these alterations should normally be covered under the landlord's policy. Any Category B items (see Section 3.4.7 for a definition) which will not be reinstated on expiry should also normally come under this policy. Any item that will be reinstated such as partitioning, decoration, cabling and fittings would normally be covered under the tenant's policy. It is sometimes advisable to include tenant's fixtures on the landlord's policy to avoid duplication and confusion in the event of a claim.

The landlord will be entitled under standard lease provisions to pass the insurance premiums (or a share of them in a multi-occupied building) on to the tenant. A different arrangement, such as a rent inclusive of insurance, would normally have to be negotiated at Heads of Terms stage and would affect the rest of the financial package.

The costs the landlord can recover from the tenant normally include the cost of insurance against **loss of rent** (and other outgoings) to cover the tenant's payments if the building cannot be occupied as a result of damage. In some cases **public and employer's liability insurance**, **property owner's liability** and **engineering insurance** may also be included. These different types of cover are discussed in more detail in Section 4.4.1. It is important that the lease is drafted to ensure that only reasonable and properly incurred costs can be charged on to the tenant. An express requirement to obtain competitive quotations is an additional protection, although in practice the landlord will probably need to do so if there is a proviso that costs are to be 'reasonable' as this requires the premium to be at the general market level.

Ideally, where the landlord is to insure, the lease will provide for the insurance to be in the **joint names** of the landlord and tenant. This protects the tenant in the case of landlord insolvency and eliminates any **subrogation rights** on the part of the insurers (rights to pursue the tenant for any damage caused by their negligence). A joint names policy is only justified where the tenant does not pay a market rent and thus holds a valuable interest in the property. It is normally only possible where the tenant holds the whole building and even in this situation it is often not practical where the landlord insures the building on a block policy.

In the absence of a joint names provision you should negotiate a right to have your **interest noted** on the policy. Most landlords are happy to note the tenant's interest but it is sometimes impractical where the property is insured as part of a group policy. The noting offers little legal protection but encourages the insurer to keep the tenant informed of claims proceedings, and generally insurers do not exercise subrogation rights against tenants whose interest is noted.

Where the insurance is not in joint names, the tenant should insist on the landlord agreeing to use best endeavours to obtain a **subrogation waiver**. This prevents the insurers pursuing the tenant for any damage caused by the tenant's negligence. While there is a common law precedent excluding rights of subrogation against tenants there is some doubt over whether the courts would uphold this and it is safest to obtain an express waiver.

Normally there will be a **cesser of rent** provision which suspends the tenant's responsibility for rent and other outgoings if the building is damaged and cannot be occupied until it is reinstated. This can cause tenants significant problems as it can be difficult to find alternative accommodation on a short-term basis, especially in a buoyant market. It is preferable for the tenant to have a **right to determine** the lease or a permanent cesser of rent in the event the premises will be unfit for occupation for a significant period. Where there is a rent-free period it is advisable to negotiate that the unexpired part of the period is postponed in the event of the rent cesser coming into force; without this term, the part of the rent-free period forgone could represent a significant financial loss. If you pay your rent in advance, you should negotiate for the landlord to have to reimburse any rent paid for the period after the cesser should begin.

The landlord's covenant will normally include an obligation to reinstate the premises in the event of damage where insurance proceeds are sufficient, unless the damage is due to some default by the tenant. The lease should ideally provide a **time limit** for reinstatement after which the tenant is no longer liable to perform its covenants. In some cases it may not be possible to reinstate at all, for example as a result of planning restrictions. Normally, if you hold a rack-rented lease and your fixtures and fittings are insured under your own contents policy, you will not be entitled to a share of the proceeds. If, on the other hand, you are paying the landlord a premium for the lease in return for a reduced rent, you should seek a provision entitling you to a fair proportion of the insurance proceeds reflecting the value of your interest.

Where you take an agreement for lease on a new development and **construction works** are yet to be completed, you need to ensure that the landlord or developer is adequately insured during the construction period. A joint names policy for this period is preferable.

Refer to Section 4.4 for a full discussion on insurance issues.

3.4.10 Alienation

The tenant's right to assign or underlet the premises follows similar principles to those for alterations outlined above. There is no common law restriction on the right to **assign or underlet** or otherwise part with possession, but most leases prohibit it without consent. Where the landlord's consent is required, the Landlord and Tenant Act 1927 implies a proviso that consent is not to be unreasonably withheld. Additionally, the Landlord and Tenant Act 1988 imposes a duty on the landlord to deal with an application for consent within a reasonable timescale. The tenant has a right to damages where the landlord fails to comply with this requirement. The 1988 Act also reverses the burden of proof in the event of a refusal so that the landlord has to demonstrate that he acted reasonably. Refer to Section 5.1.6 for a full explanation of the procedures.

Reasonable grounds for refusal can include those relating to the financial standing of the tenant and the proposed use of the premises. You should resist restrictions on your statutory right to consent in the alienation clause. Examples of such restrictions are prohibition of subletting at rents below the market level or the level reserved in the lease; or restrictions on the nature of the permitted assignee or subtenant by reference to their financial standing. These type of restrictions can severely prejudice your ability to dispose of the lease.

In some cases restrictions on alienation are unenforceable. For leases granted before 1 January 1996 the ability of the landlord to restrict the tenant's alienation rights is limited by the Landlord and Tenant Act 1927 which renders certain conditions on assignment and subletting invalid. Your lawyer can advise on whether a particular condition is enforceable. Leases granted after 1 January 1996 are subject to the Landlord and Tenant (Covenants) Act 1995 which allows the landlord and tenant to agree certain restrictions on assignments, for example by reference to financial ratios, types of use or types of company. Alternatively, the clause can include a general test about financial suitability subject to a requirement on the landlord to apply the test reasonably or to refer the matter to a third party adjudicator. The 1995 Act does not apply to sublettings so the 1927 Act provisions which qualify restrictions on sublettings will apply (refer to Section 3.4.3 particularly regarding guarantees by outgoing tenants).

3.4.11 Use

Leases usually restrict the use to which the tenant can put the premises. **Change of use** is not permitted unless this is expressly stated (even if it would be permitted under planning law) so without an express right the landlord will have full scope to refuse consent to change of use. Ideally you should try to include a provision that consent to change of use is not to be unreasonably withheld.

You should be careful to ensure that the user clause is not defined so narrowly as to stand as an obstacle to an assignment in the future. Where you are taking a reasonably long lease you should also bear in mind potential changes in your business which may require occupational changes. An example of restrictive drafting which has caused problems in recent years is a standard prohibition of gambling which has obstructed the use of premises for the sale of lottery tickets. A flexible user clause is important in a fast-changing world. However, at the same time you need to be cautious of very widely drawn clauses which can inflate the rental value at review.

It is common for retail premises, particularly in shopping centres, to include a **keep open clause** in the lease. This can take the form of a positive user clause or a separate obligation to keep open for trading. This is sometimes necessary where a landlord has planning obligations to keep open but is usually in an effort to protect pedestrian flows and the image of the centre. The clause can have a depressing effect on the rent at review and you should ensure that the rental package reflects the disadvantage to the tenant. You will also want to resist any disregard of the keep open clause at rent review as this will inflate the new rent payable.

In shopping centres you may find it hard to resist the clause as it is often a standard term. The obligation may be harder to resist for anchor tenants and other large units than for smaller tenants whose keeping open is not always as important for the centre. Where a keep open clause is accepted it is important to ensure that the drafting allows for temporary closure for specific reasons such as repairs and alterations, staff training and stocktaking.

It should be noted that the courts may enforce keep open clauses with an order for specific performance (forcing you to comply). The courts have been reluctant to do so in the past, but are more likely to do so where you are an anchor tenant in a shopping centre. A damages claim is more likely to be upheld but for this the landlord will have to show a financial loss. Some leases contain specific financial remedies for non-compliance and you can try to negotiate for provisions in the lease which would restrict the landlord's remedies to damages only. You may also need to consider your obligations in relation to Sunday and Bank Holiday opening.

For office leases landlords sometimes impose a restriction on the length of time the premises can be left **unoccupied**. Sometimes the tenant becomes liable for increased security, insurance and other costs and ideally such clauses should be avoided. The landlord will normally be able to recover any increased insurance premium under the insurance provisions in any case.

You should ensure that use of the premises is permitted outside normal business hours and that you are not encumbered by restrictions such as having to inform the landlord every time your staff wish to work late. Even if you are not likely to need to use the premises for extended hours now your needs may change and you must also bear in mind the needs of an assignee. An obligation to pay any additional costs resulting from **use outside hours** is normally reasonable.

When negotiating the user clauses, bear in mind that there may be restrictions in the landlord's title in relation to use. Common restrictive covenants in freehold title relate to the sale of alcohol, gambling and other 'immoral' uses. You may be subject to these restrictions even if they are not expressly stated in your lease. For short-term leases it is quite common not to investigate the landlord's title, in which case it is important that your solicitor raises obtains details of any such restrictions as part of the enquiries before contract.

3.4.12 Other rights and obligations

The tenant should consider in the planning and site selection stages what other rights or services are required. Examples can be the landlord providing standby power, 24-hour access or 24-hour services such as reception facilities and air conditioning, or restrictions on the landlord's right to make changes, such as his right to change the name of the building.

Where you cannot recover input **VAT**, or where there is a reasonable prospect of assigning to such a tenant, you will need to consider the implications of the landlord electing to charge VAT (Section 1.6). Ideally you should negotiate that the rent and other charges are expressly inclusive of VAT so that the landlord would have to pay any VAT, but this is rarely negotiable. Alternatively, you may be able to obtain an undertaking from the landlord not to elect, although this may be subject to limitations. Without some kind of restriction on the landlord's power to elect for VAT, you will be exposed to the risk of him using the threat to elect as a means of renegotiating the rent or other terms of the lease.

Where there is a risk that the property may have a problem of **contamination**, you need to ensure you are protected effectively from the cost and occupational implications of remedial works. This applies whether or not you are taking the lease of the whole of a site or building, as even where you are only taking part the landlord may be able to pass on to you the costs of remediation through the service charge or other clauses, and he may have a right to enter onto the demised area to carry out works.

The main issues to be addressed are:

- liability to undertake and pay the cost of remedying past contamination, whether or not identified at the time of the transaction, pursuant to a statutory notice or otherwise;
- obligations and cost liability for future contamination, whether caused by you or a third party;
- repairing obligations in relation to damage caused to the property by contamination;
- use of the property for potentially contaminative uses;
- rent review assumptions in relation to contamination problems or risks;
- liability to third parties and obligations to pay insurance premiums in respect of such risks;

- obligations to insure and pay the cost of insurance against liability for environmental damage.

In order to ensure you are properly protected it is important that wherever contamination is considered to be a potential problem it is investigated and raised early on in the negotiations. The matter can be complex and is potentially costly, so professional advice should always be taken.

3.5 STRUCTURE OF DOCUMENTATION

While the occupier does not need a detailed understanding of the technicalities, a broad understanding of legal documentation will ensure that you can contribute fully to the drafting process.

A **purchase** is normally documented by an agreement for sale detailing the terms of the transaction (exchange), followed by a transfer of title (completion).

For a **leasing** transaction the documentation may include all or some of the following.

- Agreement for lease. This is a contract to complete a lease at a later date and is used where there are conditions that need to be complied with prior to completion of the lease. It is usually required where the landlord or tenant is obliged to undertake works to the premises before the tenant can occupy.

    The agreement for lease has annexed to it the draft lease which will subsequently be completed. Both parties are contractually bound to complete the lease in the agreed form.
- Lease. This normally contains all the terms for possession and occupation, many of which are discussed above.
- Licence to alter. This documents the landlords consent to the fit out works and the terms on which it is granted. Detailed plans are attached.

    The licence to alter is used at rent review to define the tenant's alterations where improvements are to be disregarded under the rent review clause (Section 4.2). It is used on determination of the lease (normally expiry or break option) to establish the reinstatement obligations (Section 5.3).

    Where detailed plans have not been finalized and agreed with the landlord, which is often the case, it is common to agree a draft licence to alter at the time of the transaction, and to annex this to the lease or agreement for lease. The plans are approved during the early part of the fit-out process and **As Built plans** (those prepared on completion of the works and showing the actual alterations carried out) are included in the licence to alter when it is completed. The danger of this arrangement is that it could give the landlord full scope to reject the proposed works or to require amendments (depending on the exact wording of the documentation). The tenant has greater protection where all plans are approved prior to contract and annexed to a completed licence. If this is not possible as many as possible of the plans should be approved.

- Side letters. In some cases it is necessary for the parties to hold a separate document agreeing one or more of the terms. This is usually needed where a personal right is agreed as it binds only the original parties. In some cases a comfort letter is written by one party containing a statement of intent which is not normally contractually binding.

This list of documents is not comprehensive and in some cases other deeds and licences will be required. It is also worth bearing in mind when dealing with legal documentation that legal wording may not always have the meaning you would naturally ascribe to it. There may be for example implied terms or special rules of construction. If you wish to explore the construction of lease in more detail to help you in dealing with the legal negotiations, there is a useful section in Lewison (1993). However, your lawyers will be able to advise you on any technical aspects of the documentation.

FURTHER READING

Commercial Leases Group (1995) *Commercial Property Leases in England and Wales: Code of Practice,* RICS Books, London.

Cotton G. and Wynn-Jones, R. (1994) *Transactions: Commercial Leases,* Longman Law Tax and Finance, London.

Laidler, D. W., Bryce, A. J. and Boswall, J. R. G. (1995) *Guidance on the Sale and Transfer of Land which may be Affected by Contamination,* Construction Industry Research and Information Association (CIRIA), London.

Lewison, K.. (1993) *Drafting Business Leases,* Longman Group, London.

Lumby, S. (1991) *Investment Appraisal and Financing Decisions* (4th edition), Chapman & Hall, London.

Male, J. M. (1995) *Landlord and Tenant,* Pitman Publishing, London.

Nourse, H. O. (1990) *Managerial Real Estate,* Prentice-Hall, New Jersey.

Tan, C. (1992) Lease or buy, *Accountancy,* December.

Wainman, D. (1991) *Leasing,* Waterlow Publishers, London.

<table>
<tr><td>**4**</td><td># Managing costs and occupation</td></tr>
</table>

KEY POINTS

- Strategies for managing occupancy cost
- The legal and technical framework relating to occupation
- Protecting your rights and managing your occupational property to meet the needs of the business

Chapter 1 (Section 1.1) addresses cost at a strategic level and considers ways of defining, measuring and managing occupancy cost as part of a property strategy. There is also significant scope for controlling and reducing cost at a detailed level by adopting a rigorous approach to all cost components, taking an active role in cost management and consistently auditing and challenging costs.

Many occupiers fail to protect themselves and minimize cost because of a lack of awareness of the scope for doing so. Landlord's invoices are left unchallenged, rent reviews are negotiated ineffectively, potential rates savings are overlooked. In many cases occupiers can make a significant impact on cost in the long run by adopting a proactive approach and questioning all costs on an informed basis. The author recently encountered a situation where an occupier was saved £50 000 simply by discovering an error in the calculation of a service charge apportionment.

But cost cannot of course be considered in isolation. Managing cost is only one objective of corporate property management and it needs to be balanced with the objective of providing an efficient and productive property base for the operation of the business. This chapter addresses the occupational issues that affect management of operational property. It provides a framework for understanding these issues as a means of managing each of the components of occupancy cost and maximizing the efficiency of occupation. An understanding of the framework will enable the occupier to take an informed and proactive approach to all occupational issues.

Before you are in a position to run your operational property effectively, scrutinize charges and liaise with the landlord and other parties, it is essential that you have a clear and detailed understanding of the terms on which you occupy. The first step is to review the lease and any other relevant documents and prepare a summary of all the main provisions. The lease summary checklist shown in Appendix B can be used to provide the framework.

You may have a report on title prepared by the solicitor at the time of the transaction. This can be used to clarify the implications of particular clauses but it does not obviate the need to prepare your own summary. Preparation of the summary will refresh you memory on the transaction terms and prompt you to resolve any unclear points with your advisers. Leaving areas of ambiguity can be costly; for example, provisions for service of notices can sometimes be unclear and if not resolved can result in your losing important rights,

A clear and detailed understanding of the lease is essential.

such as the right to review the rent or break the lease. The summary will provide a useful reference point for quick checking in the future. Copies of the summary together with the documents should be placed in an easily accessible place on all relevant files and a diary system should be used for all important dates.

Having familiarized yourself with the documentation, you can turn to each of the components of occupancy cost. These are discussed below but refer also to Chapter 3 where many of the issues are discussed in the context of transaction structuring.

4.1 RENT

The efficiency of space usage and the rental transaction negotiated determine the rental obligations into the long term. Chapter 2 covers the acquisition process and by adopting the structured approach advocated the occupier should be in a position to achieve the best rental structure available in the market having regard to business needs. Chapter 3 discusses the important transactional issues affecting rental costs, including the transaction structure, floor area measurement, the rental rate, and incentives such as rent-free periods and capital contributions.

Rental costs are largely determined at acquisition stage but there is some scope to alter them at other times, in particular at rent review and by disposing of properties in whole or in part.

Some tenants are not aware of the powerful remedies available to a landlord for non-payment of rent. These are much stronger than those available to other creditors who usually have to rely on court proceedings and other protracted remedies to obtain payment. In addition you should remember that your landlord may be required in future to provide references and a poor payment record may prejudice your ability to take premises. Refer to Section 4.13 below where remedies for breach of covenant generally are discussed in detail.

The occupier should have an understanding of the legal framework affecting the payment of rent to determine when it is or is not appropriate to withhold payment. In general the landlord is in a very strong position when it comes to enforcing rental covenants as a result of his right to levy **distress** (i.e. use bailiffs). A tenant should be very cautious about withholding rent and can only do so legally in certain unusual situations. You may have a period of grace in which to pay the rent but this may only relate to the calculation of interest and may not prevent the landlord taking other steps immediately – you should check your lease on this point. As soon as the rent is overdue (the day following the rent payment date), the landlord can use bailiffs to remove any saleable goods held on the premises, although he may precede this with a 'letter before action' giving seven days in which to pay.

You should note that in addition to the remedy of distress the landlord can **forfeit** the lease for non-payment of rent. In most situations the landlord can do this without a court order and can simply take possession by **peaceable re-entry** if there is no one on the premises. The court may grant you **relief from forfeiture** but this will depend on your ability and willingness to pay the rent in future.

It is important that you do not allow rent arrears to accumulate or ignore notices relating to legal action. If you are having problems paying, you should keep in regular contact with the landlord, keep him informed of your financial position and take legal advice immediately proceedings are threatened. Many landlords will accept instalment payments for short periods and in some cases will grant rent concessions where this is the only practical option, although the state of the market and chances of finding another tenant will affect the landlord's level of tolerance. Provide the landlord with as much information as possible, such as management accounts and cashflow projections. This is no time to withhold information; the landlord will not grant any concessions unless he can see it will be in his best interest in the long run and most landlords will respect a wish that the information be treated confidentially.

If you are insolvent or close to insolvency you may be in a strong position to agree a settlement with the landlord as he may be unwilling to litigate when the chances of success are small. The position will be affected by the existence of original tenants or guarantors with a continuing liability and the landlord may choose to pursue these parties instead of you. This may result in a long running dispute and possibly litigation. The landlord's ability to pursue an original tenant will depend on whether the privity of contract principle applies and this in turn depends on whether the lease (or agreement for lease) was granted before or after 1 January 1996. All new leases granted after 1 January 1996 will be governed by the rules contained in the Landlord and Tenant (Covenants) Act 1995 which limit the application of the privity of contract principle, and in many cases the landlord will not be able to claim against original tenants (Section 3.4.3).

Most modern leases contain a clause enabling the landlord to charge **interest** at a penal rate (usually 2-4% above bank base rates) on rent and other charges paid late. Sometimes, there will be a period of grace, usually between ten and 21 days, during which interest is not charged. However, you should read the lease carefully on this point as many do not allow the landlord to charge interest until this period has elapsed but once it has they can charge from the due date. The landlord cannot use distress and forfeiture to enforce payment of interest unless the charge is expressly reserved as rent. As for all payments, you should always check interest calculations This is relatively quick and easy using a spreadsheet package, although specialist bolt-on packages are available which make it quicker. The landlord should provide details of the calculation or at least the interest rates used and amounts overdue.

4.2 RENT REVIEW

4.2.1 Procedures

The procedures for implementation of a rent review are governed by the terms of the lease. Leases vary quite significantly on this and it is essential that you familiarize yourself with the details well before the rent review date. Leaving the matter until the last minute exposes you to the risk that you will miss a notice date and fail to protect yourself. Additionally, you will want to budget for changes in the rental liability. If there are any complex legal or valuation issues it will be helpful to be in a position to plan for these and to address them in good time. And, in the rare situation that the review is not **upwards only** and rental values have fallen below the passing rent, it may be necessary for you to initiate the rent review. It is advisable to refer to the lease at least two years prior to the review date and make a note of any deadlines and relevant dates as well as clarifying the mechanisms for review generally and the valuation basis. If you have any queries at this stage you should clear them up with a solicitor or chartered surveyor.

Some leases provide relatively detailed procedures and deadlines for initiating and agreeing the review. Other leases leave the issue open and simply require the parties to negotiate, with rights to refer the matter to a third party if agreement cannot be reached. Where there is a deadline for service of a notice or other action, this is only enforceable where there is an express statement that **time is of the essence** or where it is implied, as in the following cases:

- if the clause contains an expression showing that time limits are intended to be obligatory, such as the wording that a notice is to be served by a date 'but not otherwise';
- if the clause states the consequence of failure to comply with a time stipulation, for example, where the clause states that the provision will be void and of no effect if the time limit is not complied with;
- if other lease provisions show that the rent review timetable was meant to be strict, such as an interrelationship with the date for service of a break notice.

This issue is complex and the interpretation of the time provisions requires legal knowledge. Where there is any doubt about how a time provision should be interpreted the occupier should take legal advice.

The rent review will normally be initiated by the landlord's service of a notice requiring an uplift in rent to a particular figure. This figure will have been chosen by the landlord as an upper figure giving scope for negotiation. Sometimes the lease will require the tenant to serve a counternotice and failure to do so within the specified period may result in the landlord's rental figure becoming the new rent. At this stage it is advisable to turn to a professional adviser, normally a chartered surveyor, if you have not already done so.

Your surveyor will need to see a copy of the lease and any other relevant documents to establish the valuation basis for the review. If you or a previous tenant have made alterations to the premises these will usually have to be disregarded for the valuation and the surveyor will need plans of the works together with relevant legal documentation, such as licences to alter. It is advisable to put the surveyor in touch with your solicitor so that he can make him aware of any relevant legal issues and so the surveyor can clarify any legal points as they arise.

The surveyor will normally inspect and measure the premises. He will then prepare a report detailing the important issues and providing an estimate of the new rent (not normally a *Red Book* valuation – see Section 1.4). He should then respond to the landlord and undertake negotiations on your behalf. The first step in the negotiations is normally to agree the floor areas. Sometimes this requires a joint inspection by the landlord's and tenant's surveyors. The surveyors will then discuss the legal issues that have a bearing on value (see below) and evidence of comparable transactions which are used to establish the market rate. Each surveyor will have comparables which support their valuation; each has the task of demonstrating why his comparables are relevant and the other's are not.

During the negotiations one party may make an offer to settle below the figure they believe they can support in order to achieve a quick resolution. If they would not wish to be held to this figure in the event of a third party determination or other legal proceedings they can mark the correspondence '**without prejudice**'. The letter cannot be used by the other party later as evidence of the rental value. A variation on this is to mark the correspondence '**without prejudice save as to costs**'. This is known as a **Calderbank letter** after a case in which it was used. This means the letter cannot be used in proceedings as evidence in relation to the valuation but the attempt to come to a resolution and curtail the legal proceedings can be taken into account by a third party in making a costs award. Using one of these procedures early in the negotiations may help to achieve an early settlement or to reduce your exposure on costs.

4.2.2 Valuation issues

The valuation is based on market evidence which has to be adapted to reflect the terms and other factors relevant to the property concerned. Chapter 1 explains the process of arriving at a rental value by adapting comparables to reflect the attributes of the subject property. This is the process that is used at rent review but it is complicated by the fact that the letting is fictitious and various artificial assumptions have to be adopted to make the valuation practical and fair. The review clause defines the assumed '**hypothetical lease**' which comprises the terms to be assumed, together with other items to be assumed or disregarded. The hypothetical lease is normally the actual lease subject to certain exceptions set out in the review clause.

This section highlights some of the valuation issues important for a tenant involved in negotiating a review. The discussion is by no means comprehensive – refer to a landlord and tenant textbook for a more detailed explanation of this complex subject.

The general principle to bear in mind is that where the assumed situation deviates from that reflected in the comparable evidence an adjustment needs to be made. Where the hypothetical lease is more restricted than the average modern lease, it will have a depressing effect on rent.

More specifically, the following are some of the important valuation issues often encountered.

- Presumption of reality. Except where the review clause expressly or necessarily requires otherwise, the review will generally be on the basis of the terms actually subsisting between the parties. Wherever there is ambiguity the courts will tend to interpret clauses to reflect the actual situation.
- User. The value of the premises is dependent on the use permitted. Where certain uses are prohibited, as is normally the case, these cannot be taken into account in assessing value. For example, where premises have a showroom use but would be more valuable as offices the office value cannot be taken into account if this use is not permitted in the hypothetical lease. If, as is usual, the lease prohibits uses which would contravene planning or other legislation, the valuation cannot directly reflect uses which are not permitted but can reflect 'hope value' for a change of use if planning permission would probably be granted.

  An unusually narrow user clause coupled with a prohibition on change of use will have a depressing effect on rent. A personal user clause (one limiting the use to the original tenant) will be interpreted as an open user clause of an unassignable lease; the increase in value attributable to the open user clause will often be counterbalanced by the decrease in value resulting from the lease being unassignable.
- Alienation. It is standard in modern leases for subletting and assignment to be permitted subject to landlord's consent which is not to be unreasonably withheld (Chapter 3). An absolute prohibition against alienation or other non-standard restrictions will usually have a depressing effect on value.
- Review period. Most modern leases provide for rent reviews at five-yearly intervals. Any different period between reviews will have an effect on value. Generally, a shorter period will reduce the rent at review because increases in rental value will filter through to the landlord more quickly, while a longer period will increase the rent to reflect the longer period before the rent can catch up with rises in rental value. The extent of the valuation impact depends on the perceptions of the market of likely rental growth, and in recessionary times the impact may be very small.
- Term. The notional term, often the original term or the unexpired residue, can have a significant impact on the rental value. In a buoyant market longer

terms are often preferred by tenants and command higher rents. In recent years tenants have been reluctant to commit to long terms and in general terms of five to ten years have been preferred.

Where the notional term is not expressly defined in the rent review clause it will usually be held to be the unexpired residue. The same interpretation is made of the wording 'for a term equivalent to the term hereby granted' or similar expression, where the term is assumed to be of the length granted by the lease commencing at the actual lease commencement date.

- Repair. Even if there is no express provision, any failure by the tenant to comply with repairing obligations will normally be disregarded.
- Fit out and incentives. Most modern review clauses include an assumption that the premises are fitted out. The aim is to ensure that the tenant, who would normally have received a rent-free period for fit out at the start of the lease, does not receive a repeat benefit. The standard disregard of the tenant's improvements means that any value added to the premises as a result of the fit out itself is not reflected in the rent; only the rent-free period normally given for fit out at the beginning of a lease is extracted from the comparable evidence to bring them onto the fitted-out basis.

In the current market other types of incentive have become common. Rent-free periods, capital contributions and stepped rents are now relatively standard. The valuation at review should reflect incentives given in the market and to ignore these would unfairly increase the rent, in some cases by a significant amount. However, some leases are worded to disregard all rent-free periods, not just those relating to fit out. In this situation the **headline rent** and not the **effective rent** will apply at review.

The courts are generally inclined wherever possible to uphold an interpretation of the lease allowing concessions to be reflected. Three cases recently decided in the House of Lords upheld this principle and decided for the tenants. However, the court can only take this approach where it can be justified by the wording of the rent review clause. Your surveyor or solicitor will be able to advise whether the lease provides scope for the landlord to obtain a headline rent.

4.2.3 Third party determinations

(a) Arbitrator or expert?

Most rent review clauses will contain provision for determination, in absence of agreement between the parties by an independent surveyor. The clause normally states whether the surveyor is to act as an arbitrator or expert. In cases where this is not clear, some guidance may be obtained from the wording. For example, reference to the Arbitration Acts will imply an arbitrator. Where the wording is unclear the professional advisers will normally be able to determine which type of third party is to be appointed. In some cases the landlord and/or tenant may have a right to choose whether to appoint an expert or arbitrator. In this situation, professional advice should be taken as the appropriate choice will depend on the circumstances.

An arbitrator is obliged to decide between the submissions of the parties but an expert determines the rent according to his own professional view and, unless directed otherwise, may not base it on the evidence or arguments of the parties. More specifically, the following principal differences apply.

- An arbitrator will request submissions from both parties and must make his decision on the basis of the evidence presented. An arbitrator's award will normally lie between the submitted figures of the parties. An expert may request submissions from the parties but will also undertake his own investigations and will use his own professional knowledge and skill.
- The conduct of an arbitration is bound by the Arbitration Acts and the rules of evidence which would apply in court. The expert is not bound by procedural rules and cannot order discovery of documents.
- There is a limited right of appeal against an arbitrator's decision. There is no right of appeal against an expert's decision but he has a professional duty of care to the parties and can be liable in negligence.

(b) Procedures

The review clause normally provides for the parties to agree on a particular third party with the option to apply to the President of the RICS or other body for an appointment, which is normally the route adopted.

On appointment, the third party will normally prepare **directions** which set out the procedure to be adopted, deadlines for submissions, his fees and the arrangements for the payment of costs. The usual procedure is for the landlord and tenant to prepare submissions by a certain date and then each to prepare a counter submission in response. Comparable evidence normally has to be confirmed in writing by the parties to the transactions concerned. The third party will normally wish to inspect some of the comparable properties. In some cases a hearing is necessary. An arbitrator may have to provide a **reasoned award** if requested, whereas an expert will be more likely to refuse. Where a **point of law** is in question the arbitrator may determine this or provide two alternative figures pending a legal determination of the point.

(c) Costs

As part of the determination the third party will normally give a direction on the payment of costs. An arbitrator has absolute discretion on costs. An expert may have this right conferred by the lease or the lease may specify how costs are to be split. Otherwise he can only deal with liability for his own fees and the parties will have to bear their own costs. Where there is an award on costs, it will normally take into account how close the parties' submissions were to the rent determined; if one party argued for a rent close to the final figure they may only pay a small proportion of the costs. Additionally the third party will take into account the efforts made to mitigate costs, so where an offer to settle close to the final figure was made and is admissible (for example a Calderbank letter – see above), costs should be awarded against the party who did not accept the offer.

4.2.4 Professional advisers

The rent settlement at review will determine the rental payments normally for at least five years and may affect subsequent rent reviews by setting the base figure for increases. The importance of rent review should not be overlooked and it is advisable to ensure you are properly protected by instructing an experienced and skilled rent review surveyor. The surveyor should normally have experience negotiating rent reviews in the locality and for the type of property concerned. Some surveyors specialize in particular types of property or areas. A good appreciation of the legal issues and their impact on value is very important. The RICS offer a referral service.

The fee basis may be a fixed fee, a time-based fee or a performance fee (sometimes a proportion of the difference between the expected and agreed rent, as an incentive to negotiate the lowest possible figure). A fixed fee related to the current or new rent is common. Be sure to obtain more than one quotation, but do not choose the surveyor purely on the basis of cost: any saving made on fees is likely to be small in comparison with the rent negotiated.

In some cases it is also necessary to instruct a solicitor. A solicitor who has knowledge of the lease can assist the surveyor as and when necessary on interpretation of the rent review and other lease clauses. If the matter cannot be agreed between the parties and a third party procedure is needed more extensive legal involvement may be necessary. In some cases a surveyor can deal with a third party procedure without legal involvement but the landlord will normally have legal advice on hand and you may put yourself at a disadvantage if you do not have access to such advice.

4.3 SERVICE CHARGE

4.3.1 Operation

The way a particular service charge is operated depends on the circumstances, the type of property and the lease. The landlord will normally prepare a **budget** of expenditure prior to or early in the **service charge year**. This will form the basis of **on account charges** (normally quarterly) made by the tenant. The landlord should monitor expenditure over the year and adjust the quarterly charges if necessary. Soon after the service charge year end, the landlord will reconcile the expenditure and prepare a **service charge account**. This may be **audited** by an independent accountant or surveyor. The landlord will make **balancing charges** or **credits** to reconcile the on account payments with the actual expenditure.

The calculation of the tenant's contribution is based on the **apportionment** percentages used. This is normally calculated on the basis of the floor areas occupied by the tenants benefiting from the service. There will normally be a number of costs shared equally between the tenants, and others to which only certain tenants contribute. Example 4.1 shows a fictitious example of a service charge apportionment calculation.

Example 4.1 Service charge apportionments for 100 The High Street, Anytown

Tenant	Floor area (sq ft)	External repairs and maintenance	Internal repairs and maintenance	Mechanical and electrical maintenance	Staff and security	Fire precautions	Energy (electricity and gas)	Cleaning	Water supply and treatment
Office A	5 000	y 25.00%	y 33.33%	y 33.33%	y 33.33%	y 25.00%	y 33.33%	y 33.33%	y 25.00%
Office B	4 000	y 20.00%	y 26.67%	y 26.67%	y 26.67%	y 20.00%	y 26.67%	y 26.67%	y 20.00%
Office C	6 000	n 0.00%	y 40.00%	y 40.00%	y 40.00%	y 30.00%	y 40.00%	y 40.00%	y 30.00%
Retail A	2 000	y 10.00%	– –	– –	–	y 10.00%	–	– –	y 10.00%
Retail B	3 000	y 15.00%	– –	– –	–	y 15.00%	–	– –	y 15.00%
Total	20 000	70%	100%	100%	100%	100%	100%	100%	100%

y tenant uses and contributes towards cost of service; – tenant does not use or contribute towards service; n tenant uses service but does not contribute towards cost.

In the example, services are used either by all tenants where all tenants benefit from the expenditure, or the office tenants only, where the service only affects office tenants. Office tenant C has a lease excluding liability for external repairs and maintenance, producing a 30% shortfall in recovery of these costs which falls to the landlord to pay.

Where the property is located on a privately owned estate or access is along private roads there may be a separate **estate charge**. This may be paid to the landlord, superior landlord or another owner.

The landlord may chose to set up a **sinking fund** for major expenditure items and in this case the service charge will include a sinking fund charge. This should normally be separately itemized. The operation of sinking funds needs to be monitored to ensure that money paid into them is not at risk. The annual sinking fund charge should be calculated to produce the sum that will be required when the repair or replacement is required, taking into account inflation. You should take legal advice to ensure that the fund is set up in such a way that the monies do not belong to the landlord in the event that he becomes insolvent.

Where you are taking an assignment of a lease from an existing tenant in a building where a sinking fund is operated, you need to ensure that the previous tenant has made the correct sinking fund contributions; otherwise you may be liable for these. Where you are disposing of a lease by an assignment, it is important to ensure that proper recognition of sinking fund payments is made and if appropriate a refund paid by the ingoing tenant, for example where the sinking fund is to be used for an improvement item from which the outgoing tenant will not benefit.

Refer to Section 1.6.2, where the VAT position relating to service charges and sinking funds is discussed.

4.3.2 Controlling costs

Service charges can represent a significant proportion of occupancy cost. Despite this, service charges are often neglected by occupiers as they are perceived as fixed and outside their control. While there is some basis for this perception – the service charge is controlled by the landlord and leases often limit the scope for challenging costs – there is often significant benefit to be gained from taking an active approach to service charge costs. In some cases large savings can be made by adopting some straightforward strategies.

Most commercial landlords are concerned to keep service charge costs down and charge tenants on a fair basis. It is often in the landlord's interest to control costs, as high service charges can prejudice marketability and where there are letting voids the landlord will be responsible for payments. However, even with the best intentions, landlords are not always focused and motivated to reduce costs they are not paying. Auditing by an independent accountant should prevent any serious errors but the auditing process does not always involve a comprehensive scrutiny of all expenditure and cost allocation.

While residential landlords are regulated by a statutory framework governing service charges, commercial landlords are not subject to such regulation. The tenant is therefore primarily dependent on the protections contained in the lease and on the willingness of the landlord to act reasonably.

A number of organizations, including the British Council of Shopping Centres, the British Property Federation and the British Retail Consortium, have together produced a Guide to Good Practice for service charges in commercial properties comprising the following guidelines.

- The owner should obtain value for money and provide services in an efficient and economic manner. Contracts should be regularly competitively tendered.
- The services should benefit the owner and occupiers.
- Service charge costs should only include enhancements of fabric, plant or equipment where this is more economic than repair. Other capital improvement and redevelopment costs should not be included.

- Apportionment of costs to occupiers should be fair, reasonable and consistent. The occupiers should not contribute to costs attributable to unlet premises or for special concessions granted to an individual occupier. The owner should bear a fair proportion of costs reflecting any use of the property for his business.
- Management fees should be reasonable for the work done.
- Owners and occupiers should communicate with each other promptly and efficiently.
- Expenditure budgets should be prepared prior to commencement of the service charge year. Accounts should be certified by an auditor and should be submitted to the occupier within six months of the end of the service charge year.

Most landlords will adopt the code of practice but this does not obviate the need for you to make your own checks. A number of strategies should be adopted to minimize costs and ensure services are run efficiently.

(a) Involvement in decision making

You should develop a close relationship with the landlord and get fully involved in the planning and budgeting process. The most important time to get involved is at the planning stage when cost decisions are made. Review the landlord's contract procedures to ensure contracts are competitively priced. Hold regular meetings to discuss budget updates and planned expenditure and ensure that you are consulted in decision making. The services are run for the benefit of the tenants and most landlords are receptive to tenants' preferences about the range and level of services.

Spending the time getting involved in the running of the service charge will not only bear direct results by highlighting problems, but will also help by focusing the landlord on the important issues. Initially, the landlord may be reluctant to involve you fully, especially if other tenants are not taking the same approach. In this situation, explain the importance of the services to your occupation and the significance of the service charge cost. In the long run, as a result of your interest, the landlord will become more focused on tenants' concerns. He will automatically keep you informed and you should be able to reduce the time spent on the matter.

(b) Scrutinizing charges

Both at the budgeting and reconciliation stage you should rigorously scrutinize all the charges. In many cases the landlord will not provide full details of the calculations and you will need to request these.

- Expenditure. Look briefly at every item of expenditure to determine whether it has been charged fairly and in accordance with the lease. This will involve inspecting invoices, which is time consuming but worthwhile. It is often permitted expressly by the lease but where it is not a reasonable landlord will normally give you access to them.

Carefully read the service charge provisions in the lease to ensure you are familiar with exactly what can and cannot be charged. Has the landlord charged any costs to the service charge which should properly be his own business cost? Are there any items which are not common costs and should have been charged to one particular tenant? Have all invoices been charged to the correct service charge code? Is there any item which goes beyond repair and should be paid for by the landlord, such as an improvement, or by the insurers where the repair relates to damage caused by an insured risk?

Look at the price for each item and consider whether it seems reasonable. If something looks very uncompetitive, check with the landlord whether it was competitively tendered. If not, speak to some contractors to obtain an indication of the going rate. Even if there is little you can do to resist the charge this year, the process may prevent the problem happening again.

Benchmark your service charge costs against other comparable properties (Section 1.1.2(d)). The *OSCAR* survey is produced by surveyors Jones Lang Wootton (annual) and you can obtain it from them or a business library. The figures only relate to offices but give a starting point for analysing costs for other types of property. Talk to other occupiers and exchange information. Other benchmarking information sources are provided in Appendix C. If you find that your service charge is high compared to others, raise this with the landlord and discuss the reasons with him.

Check the landlord's policy on interest to ensure this is fair and consistent. Where the on-account service charge income is not sufficient to fund expenditure at any point in the service charge year it may be necessary for the landlord to fund the shortfall. It is normally reasonable for the landlord to charge interest in this situation as long as interest earned on surplus service charge funds also accrues to the service charge account.

The landlord should keep a bank account for service charge monies separate from his business accounts. Ideally, there should be a separate account for each property, but for small properties this is often not practical.

- Allocation. Next you should turn to the apportionments and scrutinize the landlord's calculations. Do not assume that his or his managing agents' calculations are correct, however sophisticated their spreadsheets or accounting packages. Prepare your own spreadsheet to calculate apportionments or check the landlord's maths carefully. Consider which tenants are contributing to each item of expenditure and decide whether this is a fair allocation.

Look carefully at the lease and ensure that the landlord has taken account of any exclusions from your liability and in other respects has only charged recoverable expenditure. If you are in any doubt ask your solicitor, surveyor or accountant to advise.

These strategies can be time consuming but are usually worth the effort as there are potentially large savings to be made. A 10% saving on a service charge of £6 per sq ft for premises of 10 000 sq ft represents £6000 in only one year and in many cases the saving will be repeated over several years. In some cases much larger savings can be made, for example where a calculation error has resulted in a significant overcharge or where a large expenditure item is not properly chargeable. The process requires an investment of time in the early stages but in the long run it may change the landlord's approach to service charge management and you should be able to significantly reduce the time spent.

4.4 INSURANCE

The management of risk is an important business management function. The issues involved fall into three categories:

- identifying and quantifying the exposure;
- eliminating or reducing risk;
- transferring risk to a third party through insurance.

The first of these does not fall within the scope of this chapter but is considered briefly in Section 1.2.1(b) in the context of decision making. This section focuses primarily on the third category as it relates to occupational property. The second category will be considered briefly under risk management below (Section 4.4.3).

4.4.1 Types of property insurance

The following is not an exhaustive list but outlines the main types of insurance cover placed in respect of commercial property.

(a) Buildings insurance

This covers the cost of reinstating or repairing the building arising from damage or destruction due to fire, flood, storm, explosion, malicious damage and other related risks. Most modern policies are on an 'all risks' basis, subject in some cases to certain exceptions or limitations on cover.

The policy is normally placed by the landlord but an occupier will be responsible for building insurance where they own the freehold or in some cases where they own a long leasehold. In some other cases the landlord will allow a tenant to insure the building, for example where the tenant is an insurance company or has access to special rates, but landlords are generally unwilling to relinquish control over buildings insurance. Where they do, they will usually take a close interest in the cover and will dictate the terms of the policy.

The **declared value** of the building is established by undertaking a **replacement cost valuation**. This should be prepared by a qualified surveyor, normally a building surveyor. The surveyor calculates the total replacement cost of the building comprising:

- the cost of replacing fabric and services (including foundations, landlord's fixtures and fittings, all walls, gates, fences, gangways and bridges, landlord's plant and equipment);
- professional fees (usually assessed at 12.5–15%);
- demolition and debris removal costs (usually assessed at 5–7.5%);
- any costs relating to compliance with statutory requirements including planning, health and safety and building regulations;
- VAT on the rebuilding cost where this would not be recoverable by the party with the obligation to reinstate.

There should not be a deduction for depreciation, unless the property is insured on an **indemnity basis**, which is normally only where it is in a dilapidated condition. In some cases tenants' fixtures and fittings can be included but these are normally insured by the tenants under their contents policies (see below).

Close liaison between the valuer and the insurance broker is essential to ensure that the valuation is prepared on the correct basis. A common problem is for a surveyor to misunderstand the split of insurance responsibilities between the landlord and tenant (see below) and overstate or understate the replacement cost.

It is usual to undertake full valuations every five years unless there are changes to the building which justify them more frequently. The valuation also needs to be reviewed annually, normally by a qualified building surveyor, to update costs for inflation. This is normally a simple indexation exercise, but can involve more complex adjustments.

The **sum insured** normally comprises the declared value at the commencement of the insurance period (sometimes referred to as a **day one basis**) together with an **inflation provision**, usually 25–50%, to cover inflation during the insurance year and the reinstatement period. The inflation provision needs to be high enough to reflect the potential for a long reinstatement period, for example where planning restrictions delay rebuilding.

Obsolete and period buildings need to be considered in detail to determine the sum to be insured. In some cases these buildings would not be reinstated in their existing style and could be replaced at a lower cost. This needs to take into account any legal regulations governing reinstatement, in particular where the building is listed or in a conservation area.

At the end of 1992 insurers withdrew cover for damage caused by terrorist activity (specifically fire or explosion resulting directly or indirectly from terrorism and in Northern Ireland additional risks including riot and civil commotion). The majority of insurers include a small amount (£100 000) of terrorism cover within the buildings policy at no extra cost. Additional cover is provided under a government backed scheme known as Pool Reinsurance Company (known as 'Pool Re') and a separate premium is charged for this cover.

One particular problem with the Pool Re scheme is that it requires the property owner's entire portfolio to be covered, even if only one property is in a high-risk area. A private sector scheme run by the British Insurance and Investment Brokers' Association has the advantage of not requiring all the properties in a portfolio to be covered and can offer savings.

(b) Loss of rent insurance

A tenant will normally have a **cesser of rent or determination provision** in the lease terminating or suspending rent, service charge and other payments in the event that occupation is not possible following damage by an insured risk. It is standard practice for the landlord to insure these payments and for the tenant to reimburse the premium. The landlord normally calculates the total sum insured by reference to current rent and service charge levels and any rent reviews or likely cost increases within the insurance year.

The **indemnity period** is often stated in the lease but can be at the landlord's discretion. It is normally three years, but five years is becoming common. The indemnity period should reflect the estimated reinstatement period. Where the tenant has a determination option exercisable in the event of major damage the insurer will normally pay out for the full indemnity period (if the premises are not reinstated and relet), as long as the landlord has previously disclosed the determination provision to the insurers.

(c) Contents insurance

The tenant should normally insure contents under their own policy. This can cover furniture, furnishings and equipment and tenant's fixtures and fittings such as partitioning, doors and decorative fittings. Where the landlord places buildings insurance the items covered under a tenant's contents policy will depend on the split of insurance responsibilities between the landlord and the tenant (Section 4.4.2).

(d) Engineering insurance (mechanical breakdown)

It is common to insure mechanical and electrical plant against the cost of repair or replacement if it suffers sudden and unexpected breakdown. The policy usually includes periodic inspections undertaken by the insurers engineers on lifts, pressurized plant and some other equipment where statutory inspections are required. It is not usual to cover plant which is not affected by the statutory regulations. The premiums charged for this cover are generally low and where plant must be inspected this is often a cost-effective way of arranging inspections.

(e) Public liability insurance

This protects against claims by third parties in the case of accidental death, personal injury and damage to property and includes the legal costs of defending such claims. It is normal for landlords to take out this cover and the premium will often be charged on to the tenants with the building insurance premium as the cost relates to the landlord's liability in his capacity as manager of the building or estate.

(f) Employer's liability insurance

Where the landlord employs staff in the management of the building or estate, it is common to insure against his employer's liability and to pass on the cost to the tenant. This is most common on large estates where staff are employed to manage the buildings and common services.

(g) Business interruption insurance

This covers the occupier's consequential loss of revenue or increased costs resulting from the interruption or interference with their occupation caused by defined risks, usually relating to material damage to the building or the termination of services such as electricity and water supply.

Where the landlord's obligations in respect of undertaking repairs and providing services are limited, business interruption insurance should be seriously considered. The cost of this insurance can however sometimes be prohibitive.

(h) Contractors' insurance

This is placed during construction works and is discussed in Section 2.4.2.

(i) Latent defects insurance

This insurance is designed to meet the cost of repairing material damage caused by defects appearing in newly constructed buildings. The policy is not normally available for older buildings, although fully refurbished buildings can sometimes be insured. Where the landlord holds this cover the cost of repairing damage will not normally fall to the service charge. Most policies however limit the definition of a defect (often only the main structure is covered) and you should clarify exactly how much protection the policy affords. The period of cover is normally ten years and the premium is usually a one-off lump sum.

An occupier taking space in a new building where the landlord has not placed inherent defects insurance should discuss whether such insurance could be placed, possibly at the tenant's cost. This is particularly relevant where the tenant is not able to negotiate a limit on his obligation to contribute to the cost of repairing defects (Chapter 3), but the cost is often prohibitive.

4.4.2 Placing insurance

(a) Insurance contracts and the principle of utmost good faith

The principle of utmost good faith is a fundamental principle of insurance contracts. It does not apply to contracts generally, and does not for example apply to leases. It imposes a positive duty on the insured to disclose all material facts and to represent them accurately. A material fact is one that would affect the 'prudent insurer' in determining whether to take on the risk and in assessing the premium. They are those facts that apply at the date of the contract, which is the date of taking out the original policy or any renewal. Any misrepresentation, such as an omission or distortion of any relevant fact, even if honest, in most cases will represent grounds for the insurer to repudiate the contract.

The result of this principle is that you should ensure that any fact which may have any effect on the insurers' risk is disclosed to the insurer, the broker or the landlord. You will invariably be reliant on the broker or landlord passing on the information to the insurer, but if you have any serious doubt about them doing this it may be appropriate to communicate directly with the insurer. The safest policy is to copy the insurer on all insurance correspondence with the broker and the landlord.

The kind of issues which are material are generally those relating to the risk of damage to the property and the size of the claim. The type of issues to disclose are:

- uses or modes of occupation that might increase the risk of damage, such as any uses involving chemicals or which may affect security;
- deterioration in the condition of the premises that may increase the risk of damage, for example where a dilapidated building suffers an increased fire risk or risk of vandalism;
- vacation of premises, which increases a number of risks, such as exposure to vandalism and can increase the extent of damage as a result of delayed discovery of problems;
- changes to the management regime that may affect security levels;
- works to the premises which may affect reinstatement costs or which may involve increased risk of damage during the construction period;
- tenancy changes, such as changes in rent and service charge payments, expiry dates and break options, where this would affect any loss of rent claim.

There is no need to disclose trivial or immaterial facts, but you should err on the side of caution when deciding what should be disclosed.

(b) The policy

As an occupier your concern will be to ensure that you have effectively covered all the appropriate risks at the most cost-effective premium. This is relatively simple where you are placing all insurance yourself; it tends to be more complicated where policies are placed by the landlord, superior landlord or another party.

As far as your own insurance policies are concerned it is advisable to clarify exactly what cover you currently hold and what additional cover is available. You can then decide whether additional policies are required or indeed whether it would be cost effective to reduce your level of cover and take on additional risk. In general, insurance is cost effective as a result of the highly competitive insurance market, but for some large companies with a low cost of capital it can be cost effective to fund losses from revenue or from an insurance subsidiary.

As a general point, when negotiating any property transaction it is often worth considering the insurance route as an alternative to negotiating terms in the transaction, although this can often be uneconomic. An example is the use of business interruption or inherent defects insurance where your risk cannot be eliminated or mitigated in lease terms negotiated with the landlord.

The administration of insurance policies is an important function and the risk associated with poor administration is finding yourself in the event of a claim with an invalid or inadequate policy. You should ensure that someone with an understanding of insurance has responsibility for overseeing insurance cover, liaising with brokers and insurers, and administering claims. Policies and other documents should be reviewed and kept safely. All material facts must be communicated to the insurance company or broker expeditiously. If in doubt take professional advice; the consequences of not doing so could be severe.

It is important to review policies and levels of cover regularly. Many occupiers renew policies annually on the same basis and omit to check that the levels and types of cover are still appropriate. Contents cover must reflect acquisitions and disposals of equipment and any alterations to the space.

When some elements of the insurance cover are placed by the landlord, it is essential to ensure that there is no duplication or gap in cover. The **split of responsibilities** between the landlord's and tenant's policies should normally be agreed at the time of acquisition and may be specified in the lease, but if there is any doubt about who is covering what, this should be resolved and confirmed in writing. This problem is most common where the tenant has undertaken an extensive fit out. Refer to Section 3.4.9 where this issue is discussed further.

Monitoring the landlord's insurance cover should not stop at confirming the landlord/tenant split. Many tenants rely on the landlord to maintain appropriate cover, which in most cases they do. But it is worth taking an interest, obtaining copies of the landlords policy documents and reviewing the level of cover. In particular, it is advisable to review the valuation, preferably with the advice of a building surveyor to confirm that the **sum insured** is adequate. The potential consequence of not checking these points is finding that the landlord's cover is inadequate once it is too late; in that situation the fact that the landlord is in breach of his lease obligations will be cold comfort if he does not have the resources to make up the cost.

It is important to ensure that your interest in the property is noted in some form on the policy and particularly that the insurers waive any rights of subrogation against you. Case law indicates that insurers do not have subrogation rights against tenants (they cannot pursue the tenant for any damage negligently caused by them). However, the legal position is not entirely clear and it is advisable to obtain an express **subrogation waiver**, which most insurers are prepared to provide. The most advantageous arrangement is to take out the policy in the **joint names** of the landlord and tenant. This is worth pressing for where you are the only tenant in a single-let building but is generally not practical in a multi-let building. Any monies paid out by the insurers will be paid in joint names and in the event of the insolvency of the landlord the claims proceeds do not form part of the assets of the company to be shared amongst creditors. The insurer does not have any subrogation rights where the policy is in joint names.

Noting your interest on the policy confers no rights as such but can be helpful in practice as the insurer will generally keep you informed of large claims proceedings. It is a minimum protection where the insurance is not in joint names and insurers will not normally exercise subrogation rights against tenants whose interest is noted, but it is advisable to obtain an express **subrogation waiver** from the insurer. This is sometimes a standard term in the policy.

(c) Premiums

As far as your own policies are concerned, obtaining a competitive premium is relatively simple. It is always advisable to obtain a number of quotations and it is important to ensure that the level of cover is cost effective. Many large corporates are able to obtain competitive rates on block policies. It is often worth talking to the insurer or broker about increasing the size of any **excess** or excluding certain risks (subsidence, for example can be very expensive). Where the landlord insures, it is normal for the tenant to bear the cost of the excess in the event of a claim as they receive the benefit of the reduced premium. Implementation of a **risk management** initiative may also reduce premiums in the long run (Section 4.4.3 below).

Landlords are often able to obtain very competitive premiums where they hold a large property portfolio, and it is sometimes in the tenant's interest to place as much as possible of the insurance cover through the landlord. In some cases however there can be problems where the landlord is not concerned to ensure that premiums are competitive. This is most common where the landlord is an insurance company or has links with a particular insurer and places insurance routinely in house or with the associated company.

Where the tenant is concerned that an insurance premium is not competitive the protection afforded by the common law is limited and they will primarily be reliant on the wording in the lease. If the lease provides that the premium must be 'reasonable' the tenant will only be liable to pay a contribution in line with market rates. This will not necessarily mean that the premium must be the very lowest available where this is not representative of the market level generally, but it does protect the tenant from a premium above the general market level. There is an implication that the landlord is obliged to shop around.

Where the lease only requires the landlord to incur the cost 'properly' this has been interpreted in recent case law to mean only that the contract must be at arm's length. Where the insurance is placed as an arm's length transaction the landlord does not have to choose a policy in line with the market. Where the contract is not at arm's length the landlord does have to ensure that the premium is in line with the market with the implication of a requirement to shop around. This provides some protection to tenants faced with landlords placing insurance in house or with associated companies.

Where the lease does not expressly limit the level of premium by use of words such as 'reasonable' or 'proper' there is a possibility that the courts will imply an obligation to obtain a competitive rate, but case law on this is ambiguous.

It is common for landlords to obtain a **commission** from the insurer in relation to the insurance policy. Where the landlord acts in the role of broker, administering policies and claims, and the policy has been placed competitively, a commission arrangement is usually reasonable. However, if a there is a separate broker also paid a commission you should dispute the arrangement with the landlord.

4.4.3 Risk management

Implementation of a risk management scheme will minimize the chances of damage and reduce insurance premiums in the long term. Most insurers will run a risk management scheme including annual or more frequent inspections by a surveyor with recommendations for reducing the risk. Recommendations should normally be implemented and the inspections should not be viewed negatively but should be seen as a beneficial service. Most leases impose liability on you if the landlord's policy is vitiated as a result your occupation, so it is important that you comply with the insurers' requirements.

You should ensure that the landlord or his insurer does have in place a comprehensive risk management plan and that the landlord is effective at implementing recommendations. Where you are placing insurance yourself it is advisable to discuss the risk management approach with the insurer to ensure that an effective scheme is in place. You should take a proactive role in considering ways of reducing risk and not rely entirely on the insurer's surveyor.

The main property hazard is fire, although other risks, particularly terrorist damage, malicious damage and flood damage are also important. The main causes of fire are malicious arson, electrical damage and contractors' carelessness. These should be kept in mind when considering risk improvement strategies. Particular attention should be paid to high risk properties, such as vacant or dilapidated property and premises occupied for hazardous uses, such as storage or use of chemicals.

The first step in risk management is to formulate an effective **contingency plan** in case of damage to ensure that safety is maintained and losses are minimized. All staff should be trained in implementing the plan. It is often advisable to instruct a **fire safety consultant** to review the building, advise on emergency procedures and train staff. This service is often very cost effective. The consultants should follow up annually with safety audits. Other safety consultants should be instructed in certain high risk areas, such as the City of London, where terrorist risk is substantial.

4.4.4 Claims

It is advisable to clarify claims procedures and obtain claims forms in advance to ensure that the claim can be handled smoothly. In the event of damage the first step is to determine which policies should be claimed against. In some cases a claim will involve a number of different policies and it is important that all the relevant insurers or brokers are informed immediately of the damage. The claim should be made against the policy relating to the item damaged not the policy relating to the item causing the damage. There is sometimes confusion on this point, for example where a leak from the building structure covered under the landlord's policy damages tenants' fittings covered under their contents policy.

The importance of having confirmed the landlord/tenant split is demonstrated at the time of a claim. If this has not been clarified there can be significant confusion and the claim may be prejudiced. At worst, you may find that some items are not covered on either policy.

If the claim is likely to be substantial the insurance company will appoint an **adjuster**. He will act for and be paid for by the insurer but will usually provide useful advice on the procedures for pursuing the claim. It may be advisable to appoint an **assessor** who will act on your behalf and negotiate the claim with the insurer. This cost will fall to you but for large claims it is often cost effective especially as the assessor will often work on a contingency fee basis. The insurance policy will not normally cover the assessor's fee. Beware of 'cowboy' assessors and only use a reputable firm.

Emergency works should be undertaken if they are necessary to **mitigate the loss**. You should always try to obtain a competitive price and the insurer should be informed as soon as possible. Always keep hold of quotations, invoices and correspondence. For non-urgent work you should always obtain the insurers' agreement before incurring any cost.

It may be possible to obtain a **payment on account** where the claim is significant. When the claim is finally settled you will be required to accept payment in **full and final settlement**. It is important to satisfy yourself that the figure is adequate before signing an acceptance form.

If there is significant damage to the building and you cannot occupy to carry on your business, there will normally be special provisions in the lease governing your position. Normally the rent will be suspended for the period until reinstatement. This **cesser of rent** may only apply during the period of the loss of rent insurance cover, in which case you will have to recommence paying the rent and other charges once the loss of rent cover is exhausted. Alternatively, you may have a **right to determine** the lease, in which case your obligations can be terminated fully. If you have any consequential loss policy which may cover relocation expenses or business interruption costs you must ensure you mitigate your loss by obtaining new premises speedily and cost effectively.

The landlord will normally have an obligation to apply any insurance proceeds to **reinstatement**. If the landlord is prevented from reinstating because of planning or other problems, you may be entitled to a portion of the proceeds. Normally you will rely on the lease provisions but in some cases you may have a common law entitlement to part of the payout.

4.5 REPAIRS AND MAINTENANCE

4.5.1 Legal issues

A modern commercial full repairing and insuring (FRI) lease or effective FRI lease obliges the tenant to maintain and repair the demised premises. In some cases, particularly for short leases, the landlord may have the obligation to repair, but this is uncommon. The wording of the repairing covenant varies but usually includes the obligation to put and keep in good and substantial repair or good tenantable repair.

The **meaning of repair** depends on the particular circumstances and the standard required will depend on the age, character and locality of the premises at the time the lease was granted. The tenant will normally be obliged to maintain the premises in a condition that would be acceptable to the average tenant taking the premises at that time. This will normally include works to remedy any disrepair which existed when the premises were taken, as the standard repairing covenant 'to keep in repair' will imply an obligation to put the premises into repair not just to keep them in the condition in which they were let. Repair can include renewal of subsidiary parts of the building, as long as this does not mean the landlord is given back a building wholly different from the premises as demised. Improvements are not normally included, unless they are necessary to effect a repair.

In practice, whether a particular item is considered to be repair depends on the circumstances, particularly the nature and the cause of the damage. In one case a raft supporting foundations was held to be within the meaning of repair while in another case, albeit 100 years earlier, a similar item had been held to be excluded; similarly, in one case, damp proofing a wall was excluded from repair while in another a defective damp proof course was included in the repairing obligation. If in doubt about your obligations take the advice of a lawyer and/or building surveyor.

Inherent defects are not necessarily excluded from the scope of a covenant to repair, although the tenant will not be obliged under a standard repairing covenant to give back to the landlord a wholly different thing than that originally demised. Unless inherent defects are expressly excluded you will often be obliged to repair the damage, which can be very expensive.

The general meaning of repair has to be seen in the context of the particular wording of the lease and there may be specific exclusions to the tenant's obligation. Where the landlord insures, it is normal to exclude damage caused by any of the insured risks. It is relatively common to limit the obligation to maintaining the premises in the condition they were in at the time of the letting, evidenced by a **schedule of condition**. Other exclusions may relate to patent defects, inherent defects, structural repairs or particular items such as plant and equipment (Section 3.4.7).

Where the tenant holds a lease of part of a multi-occupied building the repairing obligation will be limited by the definition of the premises and will not normally extend to any structural items. These fall to the landlord under the service charge.

The repairing clause may extend the tenant's repairing obligation by express mention of types of work that would not normally fall within the meaning of repair. Thus, renewal or improvement may be specifically included within the tenant's obligation.

The lease will often include an express obligation to **decorate and clean** at particular intervals, often five yearly internally and three yearly externally. You may need to accommodate this in your planned maintenance programme. If for any reason you do not wish to comply with the repairing obligation strictly, the landlord may choose not enforce the covenant during the lease in which case the matter can be addressed in a dilapidations claim on expiry (Section 5.3).

You should check your lease to see whether the landlord has a **right to enter** to do repairs. Many leases contain such a provision and enable the landlord to undertake repairs himself, often merely on service of a letter of notification. He will normally be able to recover the cost of the repairs from you.

Where the tenant is in breach of the repairing covenant and the lease was originally granted for a term of at least seven years and has at least three years remaining, the tenant may be able to claim relief under the **Leasehold Property**

(Repairs) Act 1938. This limits the tenant's repairing obligation to works that fulfil certain limited criteria, such as those works necessary to prevent a substantial diminution in the value of the landlord's reversion, those required by statute and those needed to prevent larger expense in the future. Where these provisions apply and the landlord takes steps to enforce the repairing covenant, you will need to follow certain procedures, including service of notices, in order to claim protection of the Act.

There are other statutory limitations on a tenant's repairing liability, including those contained in the Landlord and Tenant Act 1927 limiting the landlord's claim to the amount by which the failure to repair diminishes the value of his reversion; and Section 147 of the Law of Property Act 1925 which gives the tenant relief from certain **internal decorative repairs**. These statutory restrictions can be very significant and you should take professional advice to ensure you take advantage of them fully. If you do not follow the procedures properly the landlord may be able to enter onto the premises to undertake the works himself and recover the full cost from you, even for items that would have been excluded if you had followed the procedures required to take advantage of the legislation.

The landlord will normally have obligations to repair and maintain the structure and common areas of a multi-occupied building. He will usually be able to charge all or most of the cost to the tenants as part of the service charge. In many modern leases the service charge clause is drafted widely to extend beyond the legal definition of repair, and renewal and replacement may be expressly included within the costs that can be charged to tenants (refer to Section 5.3 – dilapidations and reinstatement).

4.5.2 Planned maintenance

Planned maintenance is preventative maintenance; it is the system whereby items are repaired or replaced before their failure would cause disruption to occupation. The programme is designed around the life expectancy of components and the maintenance and servicing frequency required to prevent disrepair occurring. Reactive maintenance often proves to be uneconomical and disruptive.

Maintenance procedure documents are usually provided by manufacturers for items of mechanical plant and equipment, such as lift motors, boilers and air conditioning units. These specify the maintenance task to be carried out and the frequency. If these tasks are not regularly and correctly carried out the manufacturers' warranties may be invalidated. Furthermore, staff safety can be compromised if essential services are not adequately maintained, particularly for emergency equipment such as fire alarms and emergency lighting.

Computer packages are available to assist in preparing planned maintenance programmes. A building surveyor can advise on developing and implementing planned maintenance systems and the facilities management books listed in Further Reading provide more detail on the subject.

4.6 ALTERATIONS AND IMPROVEMENTS

An alteration is any change affecting the structure or form of the premises. An improvement is something new for the benefit of the occupier and is viewed from the tenant's perspective. Thus, an item may be legally defined as an improvement even though it reduces the value of the landlord's interest, for example by reducing the rental value of the premises. The main legal issues which need to be addressed when considering alterations to leased premises are:

- the alterations covenants in the lease, including restrictions on the right to make alterations and requirements to obtain landlord's consent;
- procedures for obtaining consent where there is a qualified covenant;
- any rights to compensation from the landlord on expiry;
- any statutory restrictions, such as planning, building regulations and health and safety regulations.

This section will consider the first three items. Statutory regulations are discussed below in Sections 4.9, 4.10 and 4.11.

4.6.1 The lease

In the absence of an express restriction the tenant has a relatively free hand to make alterations. However, certain alterations may be prohibited under other clauses if they damage the premises, and there will normally be a requirement under the yielding up provisions to reinstate any alterations on expiry or sooner determination of the term.

The usual alterations covenant in a modern lease will prohibit alterations to principal or load bearing parts of the structure, and allow other alterations with the landlord's consent. Certain minor alterations, such as the installation of non-structural demountable partitioning may be specifically allowed without the landlord's consent but you may still be obliged to inform him about the works.

In rare cases, there is an absolute prohibition on all alterations. This can sometimes be circumvented as follows:

- Where the works fall within the definition of repair they may be permitted by the repairing covenant.
- If the tenant is under a statutory obligation to do the work you can normally apply to the court for a variation of the lease.
- You may be able to obtain a right to undertake works where an application under the compensation provisions of the Landlord and Tenant Act 1927 is made (see below). This is an often forgotten mechanism which can be used to circumvent an absolute prohibition.

4.6.2 Consent to alterations

Where alterations are permitted with landlord's consent, the tenant must apply to the landlord providing details of the proposed works. If you are permitted to make alterations with the landlord's consent the Landlord and Tenant Act 1927 implies a proviso that consent to **improvements** is not to be unreasonably withheld, even if this is not stated in the covenant. The 1927 Act severely restricts the landlord's scope to refuse consent to improvements. Reasonable grounds for refusal include a failure to obtain planning permission where this is required, any other statutory prohibitions against the work, or in some cases breach of other covenants in the lease (in which case the consent must be granted once the breach is remedied).

You should ensure that full details of the works including plans and specifications are provided in the initial approach to the landlord. It is advisable at this stage to provide an undertaking to reimburse the landlord's reasonable **costs** in dealing with the consent. The landlord will not normally proceed until he has obtained such an undertaking and to provide it quickly will avoid delay. The landlord may require an undertaking from your solicitor. Where he requires an onerous undertaking or he delays on the issue of costs you have no direct remedy and will have to rely on the court declaration procedure discussed below or take the risk of proceeding without consent.

You should ask for an estimate of the landlord's costs but you cannot expect a definite figure as it will depend on the time taken by the lawyer and other professionals. The landlord's costs for straightforward minor alterations will normally be in the range £200–600 plus VAT, usually comprising surveyors' and legal fees and possibly an additional charge by the landlord. For more extensive works or where there are legal complications the costs can be significantly higher and will relate to the time taken by the landlord's consultants to review plans, suggest amendments and inspect the works, and for the lawyers to draft appropriate documentation. Where your landlord holds a leasehold interest he will probably have to obtain the superior landlord's consent which will increase the costs. You will have little scope to challenge costs unless they are significantly out of proportion to the work involved.

The landlord can in some cases impose conditions on consent. Where the alteration diminishes the value of the landlord's interest the tenant must be given the option of paying the landlord **compensation** for the diminution in the value. If you are prepared to compensate the landlord fully for the diminution this will not be a reasonable ground for refusal. Where the works do not add to the letting value of the holding, the landlord can impose a **reinstatement** obligation. In practice, few alterations add to letting value as most suit the particular needs of the tenant and will not benefit the average tenant in the market. Where, however, the works are to base building structure or services, such as enhancement to ceilings, floors or air conditioning and would not in any way inhibit occupation, they will probably not have any prejudicial effect on letting value and you may be able to resist a reinstatement obligation.

If the landlord does not respond to the application within a reasonable timescale, or he unreasonably refuses consent or imposes unreasonable conditions, such as an unreasonable reinstatement obligation or requirement for compensation for diminution in value, the tenant can apply to the court for a **declaration** that the works are a proper improvement. This will only be granted if, amongst other things, the improvements will add to the letting value of the premises.

If you are sure of your case you can simply proceed with the works and, if the alteration would be defined as a proper improvement by the court, the landlord will have no claim against you. This is a highly risky course of action as the landlord may be able to forfeit the lease following service of a notice requiring you to remedy the breach of covenant and you may prejudice your right to compensation. You will only be granted relief from forfeiture where the works would have been allowed by the court. In a poor market where the landlord will not use the forfeiture remedy you are in a stronger position. You take the risk however of a damages award (limited to the diminution in the value of the landlord's reversion) and/or of the landlord obtaining an injunction preventing you from proceeding with the works and requiring you to reinstate the works already carried out.

If you face a refusal situation you should always take legal advice, firstly to determine whether the alteration is covered by the Landlord and Tenant Act 1927 and would be defined as a proper improvement, and secondly to ensure that the correct procedures are followed.

Where the landlord grants consent this will be documented by a **letter licence** or a **formal licence** to alter. A letter licence should only be used where the alterations are minor and there is no reinstatement obligation. The landlord may include conditions on the consent, whether or not a formal licence is used. The conditions commonly encountered are:

- an obligation to reinstate the premises to the condition they were in before the works were undertaken;
- a requirement to obtain and provide evidence of all statutory consents, including planning and building regulations approvals;
- an indemnity for all liability arising from carrying out the works or from the works themselves;
- a requirement to notify the insurers of the works, to comply with their requirements and pay any increase in premium resulting from the works;
- a deadline for completion of the works;
- an obligation to notify the landlord once the works are completed;
- inspection of the works by the landlord at the tenant's expense at the end of the works;
- an undertaking to reimburse any costs incurred by the landlord in dealing with the works or the consent.

The rent review clauses in the lease will normally contain a disregard of tenants' alterations at review. For the avoidance of doubt it is advisable, though not essential, to get a clause included in the licence confirming that any increase in rental value attributable to the works is to be disregarded for the purpose of rent review.

4.6.3 Compensation for improvements

Tenants have a right to obtain compensation from the landlord for improvements under Part 1 of the Landlord and Tenant Act 1927. This is a commonly neglected mechanism, largely because of the difficulty of demonstrating that the works add to the letting value of the premises, and because the landlord can choose to do the works himself in return for an increase in rent. Professional advice should be taken on whether the procedure is appropriate in the circumstances.

The amount of compensation is limited to the net increase in the value of the landlord's holding, and many of the works undertaken by tenants do not enhance value as they would not benefit the average tenant. However, where a tenant is undertaking a substantial fit out or other significant works at the commencement or later on in the term, there will often be items included which increase value.

Compensation is payable on expiry or sooner determination of the term, but the tenant has to have made **prior application** to the landlord, notifying him of the works. The application for consent under a qualified alterations covenant (as described above) will normally suffice as an application under Part 1 of the 1927 Act.

If the landlord objects to the tenant's application the tenant can apply to the court for a declaration that the alteration is a proper improvement as discussed above. If the declaration is granted the right to compensation will apply.

The **claim** for compensation must be made within three months of the end of the lease by, for example, a Section 25 notice terminating a contractual tenancy, or forfeiture of the lease by re-entry or a notice to quit.

Legal advice is essential in determining whether the improvement is covered by the compensation provisions, in making the application prior to the works, obtaining any court declaration and securing compensation when you exit.

4.7 OPERATIONAL SERVICES

For the purposes of this book the term facilities management is used to mean the management of operational services, ranging from property related services such as space planning and space management, building maintenance and management and project management to support services such as security, mail, catering, travel, vending, procurement and so on. Strategic portfolio management is considered to be a separate though related function and is discussed in Chapter 1.

The term has become a buzz-word in recent years, as professionals and contractors have identified a growing need from occupiers in the context of diminished demand from landlords. Many facilities managers define the term to include the full spectrum of work related to occupational property, ranging from business management services, such as accounting and marketing, advice and management related to legal issues such as leasing and employment, through to provision of operational and support services.

It is not within the scope of this book to look at facilities management in any detail; the subject could (and does) fill a book of this size several times over. Clearly, facilities management is central to corporate property management and is an area the property manager will wish to explore in detail. Bernard Williams Associates (1995) and Park (1994) cover the issues in detail.

One issue which is worth highlighting here is outsourcing the facilities management function. This is becoming relatively common, particularly for large organizations. Specialist facilities management providers can often demonstrate cost savings to be gained from outsourcing facilities management but the occupier should be cautious when evaluating such claims. As discussed in Section 1.1.2, occupancy cost can be measured in many different ways and figures can often be manipulated to support a particular claim. In many cases there are potential gains to be made from outsourcing facilities management to the right specialist, driven by a variety of factors including economies of scale and specialist skills, but any analysis used to support such claims should be scrutinized carefully.

Most facilities management providers come either from a contracting background or from a professional background at the operational end of the property market, although many have bolt-on strategic management skills. These advisers can often add significant value in the management of operational services but it is important to ensure that you instruct the most appropriate consultant to advise on each area. In particular, many strategic and leasing issues require advisers with a different range of skills, such as lawyers, accountants, management consultants and general practice surveyors.

The decision to outsource facilities management will normally be driven by three primary objectives:

- the wish to achieve cost savings by taking advantage of economies of scale and specialist skills;
- a need to enhance the quality of service to users by use of specialist providers;
- a concern to focus in-house skills and resources on the core business and avoid redirecting resources which would be used more productively in the core area.

The decision will depend on strategic business objectives and the availability of in-house resources. Broadly, the outsourcing options available relate to the range of services covered and the extent to which management and delivery are included. The main alternatives are as follows.

- Single service contractor. These companies concentrate on the delivery of one particular service, such as security or catering. They can be instructed directly but more usually as subcontractor to a main facilities management contractor.
- Packaged service contractor. These companies draw together a variety of services but usually specialize in a particular type of service, such as office services or building services. They tend to be instructed as subcontractors.
- Management contractor. These provide particular services together with a management function. You could choose to hand out the management of a particular operational function and they would provide both delivery and management.
- Total facilities management (TFM). This is full management of a range of operational services. The main management contractor will normally subcontract a number of the services.
- Management company. This is a management only service. The company manages a range of contractors delivering services on direct contract with the user.
- Consultant. An occupier may choose to manage services in-house but may need to take the advice of facilities management consultants on a continuous basis and/or on one-off projects.

4.8 USE

Most leases restrict the use for which the premises can be occupied. The user clause may refer to a specific use, such as 'commercial or professional offices', or, as in many modern leases, to a class in the Use Classes Order (Section 4.9). In some cases other definitions are adopted, such as a 'high class' use. This type of wording is usually interpreted widely and is sometimes unenforceable.

If you wish to change the use of the premises, the assistance afforded by statute is limited and your ability to do so will depend on the wording of the lease. Where the covenant is absolute the landlord is entirely free to prohibit a change of use, except in the case of certain long leases. Where the covenant takes the qualified form (it is subject to landlord's consent) and does not involve any structural alterations, the Landlord and Tenant Act 1927 provides that the landlord is not entitled to extract any fine or premium, although he can ask for compensation for damage to the value of his interest and for reimbursement of costs in dealing with the application and in practice there may be little you can do to prevent a landlord demanding payment. Unlike for alterations and alienation, the 1927 Act does not introduce a requirement that consent is not to be unreasonably withheld if this is not expressly provided and the landlord has a free hand to refuse consent. Where the change of use involves alterations there is nothing to stop the landlord taking a premium.

In considering the use of the premises and any change of use you must have regard to planning as well as lease restrictions.

4.9 PLANNING

The planning system governs both the use of property as well as any building works carried out. The occupier will need to consider planning issues when considering a change of use and when dealing with alterations to premises or any other kind of construction work. This section provides a brief explanation of the planning regime to help you work your way through the system effectively. It will assist you in determining when planning permission is required and explain the main steps in obtaining it.

Planning law and procedure can be a complex and detailed area. This section can only outline the main framework and the key issues relevant to an occupier. If you wish to know more details, refer to one of the textbooks listed in Further Reading, and if appropriate take professional advice.

4.9.1 Need for planning permission

You will require planning permission if your proposals amount to 'development', unless a relevant exception applies. **Development** is defined in the Town and Country Planning Act 1990 (Section 55) as:

1. the carrying out of building operations, engineering operations, mining operations or other operations in, on, over or under land, or
2. the making of any material change in the use of any building or other land.

As an occupier you may have specialist dealings with engineering, mining or other operations. The main issues of relevance to occupiers in general however are building operations and change of use, so these are the focus of this section.

(a) Building operations

Building operations include:

- erection of a new building;
- structural alterations or additions to buildings;
- demolition of a building in whole or part, subject to a number of exclusions;
- other construction works.

Certain works are specifically excluded, in particular, improvements, alterations, or maintenance works which do not affect the external appearance of the building.

By the '**de minimis**' principle very minor building operations, such as most decorative changes, do not fall within the definition of development and do not normally require planning permission.

The position relating to demolition is relatively complex. As a result of the large number of types of demolition excluded from the definition of development, planning control principally applies to the demolition of dwellinghouses outside conservation areas.

(b) Change of use

No slight or trivial change of use will amount to development and planning permission is only required for material changes. What amounts to a material change, however, is a question of fact and degree to be determined in the circumstances. There is clearly a change of use when the nature, rather than just the type of the use is changed, such as when a business use is changed to a residential use. There may be a dominant and ancillary use, such as retail or industrial use with ancillary offices, but if the ancillary use extends this may become a separate use in its own right triggering a need for planning permission.

The Town and Country (Use Classes) Order 1987 (UCO) specifies classes of use within which changes are not deemed to be development and do not require permission. The UCO is only relevant in relation to material changes of use. The classes listed in the UCO are summarized in Table 4.1. For example, a change from a bank to a betting shop remains within class A2 and does not need planning permission.

The UCO is not comprehensive. Certain uses do not fall within any class and are 'sui generis'. These include theatres, car showrooms, amusement centres, petrol filling stations and car hire offices.

The fact that works constitute development does not mean necessarily that planning permission is required. Certain types of development are given planning permission automatically (see below). Additionally, enforcement action cannot be taken ten years after a breach of planning control depending on the breach (see Section 4.9.4).

If you are in doubt as to whether you require permission for particular works or a change of use and wish to clarify the position, the first step is to discuss the matter informally with a planning officer at the local planning authority (the district council, unitary council or metropolitan authority). If you require confirmation that planning permission is not required for either building operations or a change of use, you can apply for a **Certificate of Lawfulness for Proposed Development**. If you wish to confirm that the existing use or presence of buildings is lawful you can apply for a **Certificate of Lawfulness for Existing Development**.

Table 4.1 Use classes

Class	Uses included
A1 – Shops	Superstores, hypermarkets, retail warehouses, hairdressers, sandwich bars, travel agents, showrooms (except car showrooms)
A2 – Financial and professional services	Banks, building societies, estate agents, betting offices
A3 – Food and drink	Restaurants, cafes, public houses, wine bars, take-aways
B1– Business	(a) offices (b) research and development (c) any industrial process that can be carried out in any residential area without detriment to the amenity of the area
B2 – General industrial	Any industrial process other than one falling within Class B1
B8 – Storage or distribution	Warehouses, distribution centres
C1 – Hotels	Hotels, boarding houses, guest houses
C2 – Residential institutions	Residential accommodation with care including hospitals, nursing homes, residential schools and colleges
C3 – Dwellinghouses	A house or flat occupied by no more than six persons living together as a single household
D1 – Non-residential institutions	Medical and health services, religious buildings, museums, libraries, art galleries, exhibition halls
D2 – Assembly and leisure	Cinemas, concert halls, bingo halls, casinos, dance halls, bowling alleys, swimming baths, gyms, and all indoor and outdoor sports, not involving motorized vehicles or firearms

(c) Permitted development

The General Permitted Development Order 1995 (GPDO) grants automatic planning permission for certain kinds of development, including:

- demolition of all buildings except dwellinghouses;
- certain works within the curtilage of a dwellinghouse, such as enlargement and improvement and the erection of freestanding buildings;
- extension or alteration of industrial and warehouse buildings.

Additionally, the Order deems permission granted for a number of changes of use, including those in Table 4.2.

Table 4.2 Permitted change of use

From	To
A1 (shops)	Mixed use: A1 (shop) plus single flat
A2 (financial and professional services)	Mixed use: A2 (financial and professional services) plus single flat
A2 (financial and professional services) with display window at ground floor level	Mixed use: A1 (shop) plus single flat
Mixed use: A1 (shop) plus single flat	A1 (shops)
Mixed use: A2 (financial and professional services) plus single flat	A2 (financial and professional services)
Mixed use: A2 (financial and professional services) with window display at ground floor level plus single flat	A1 (shops)
A3 (food and drink)	A1 (shop)
Car showroom	A1 (shop)
A3 (food and drink)	A2 (financial and professional services)
A2 (financial and professional services)	A1 (shop) with window display at ground floor level
B2 (General industrial)	B1 (business)
B8 (storage and distribution) – no more than 235 sq m of a building's floorspace can change to or from Class B8	B1 (business)
B1 (business)	B8 (storage and distribution) – no more than 235 sq m of a building's floorspace can change to or from Class B8
B2 (general industrial)	B8 (storage and distribution) – no more than 235 sq m of a building's floorspace can change to or from Class B8

The permitted development rights granted under the GPDO are qualified by conditions and exclusions. In particular, permitted development rights are curtailed in some cases, for example in National Parks, Areas of Outstanding Natural Beauty and where the Local Planning Authority has made a direction under Article 4 of the GPDO restricting particular rights.

A more relaxed regime applies in **Enterprise Zones** and **Simplified Planning Zones** (SPZ). In these areas planning permission is automatically granted for development specified in the Enterprise Zone or SPZ scheme.

4.9.2 Planning permission

Obtaining planning permission can take anything from four weeks, if the application is approved immediately, up to a year or more for a complex planning appeal. The application is determined by the **local planning authority** (LPA).

With the exception of certain special areas such as National Parks and the area covered by development corporations, such as London Docklands, the LPA is the local authority, which is the district or borough council. However, applications concerning waste disposal and minerals are dealt with differently.

Planning policy is formulated both at central and local government level with three tiers of guidance. Firstly, there is central government guidance, issued by the Department of the Environment, in the form of White Papers, Planning Policy Guidance Notes (PPGs), circulars and Regional Planning Guidance (RPGs). In rural counties strategic guidance is contained in the **Structure Plan** issued by the county council. The third tier of policy is at local level; for the rural areas this is the **Local Plan** prepared by the district or borough council. The metropolitan districts and London boroughs issue a **Unitary Development Plan** (UDP) which combines strategic and local guidance. The **development plan** is the UDP in the metropolitan districts and London boroughs and the Structure Plan and Local Plan together in the rural areas. Under the current reorganization of local government some former rural district councils and some former counties have become 'unitary authorities' and will issue a UDP for their area, as if they were a metropolitan district or London borough.

The issues to be taken into account by the authority in determining an application include:

- the development plan (the UDP or the Structure Plan and Local Plan together). Since the Planning and Compensation Act 1991, there is a presumption in favour of development proposals which are in accordance with the development plan. Where the development plan contains relevant policies the application will be determined in accordance with these, except to the extent that material considerations indicate otherwise.

  The status of the plan will be affected to some extent by how recently it has been prepared. Where either the UDP, Structure Plan or Local Plan was prepared some time ago its status may be diminished. In some cases the policies contained in new draft plans may be of greater relevance, depending on the stage they have reached in the formal adoption process;
- government guidance;
- amenity considerations, such as visual intrusion, noise, pollution and traffic;
- planning gain or advantage directly associated with the development, such as local infrastructure or facilities;
- any special planning considerations relevant where the proposed development is located in a National Park, Area of Outstanding Natural Beauty, Green Belt or other specially designated area.

Once you have determined that planning permission or other consent is required you will need to establish your chances of success. The first step is to review the development plan for the area, either the UDP or the Local and Structure Plan. From these you will see whether the building falls into any special policy zones and the considerations which will be applied in determining applications. You should ensure you consult copies of both the adopted plan and any consultation or 'deposit' draft which has not yet been formally adopted.

Before submitting an application it is usually advisable to discuss the matter with a planning officer at the LPA who can give you an indication of the LPA's likely approach to the proposed development. Always ensure you speak to the officer responsible for the particular site and one of enough seniority to ensure he will give reliable advice. For all but the most simple proposals it is important to provide details of the development in writing with plans; this will avoid a misleading response from the planning officer resulting from incomplete or inaccurate information. In some cases, a site visit is necessary. Also, to avoid any misunderstanding always ask them to confirm anything they say in writing, or do so yourself.

You should bear in mind that while the planning officer's response is of importance, you should not rely on it exclusively; in most cases the decision will be made by the councillors sitting on the planning committee and they do not always follow the advice of the planning officers. Equally, do not assume that just because the planning officer is encouraging you are certain to be granted permission. If the application is refused, you will have a right of appeal to the Secretary of State for the Environment. If you think there is support for your proposal in planning policy documents, it may be worth pursuing an application despite discouragement from the officer. Where you are not sure, professional advice should be taken (Section 4.9.7).

In some cases where you are applying for consent for erection of a new building, it is advisable to obtain **outline planning permission** only. This is particularly relevant where you are not intending to undertake the development yourself but are selling land for development and wish to sell it with the benefit of planning permission or where you may wish to change parts of the design later without having to reapply for the main permission itself. The outline permission will be granted subject to certain reserved matters, details of which have to be approved before the permission can be implemented. These may include issues relating to siting, design, landscaping, external appearance and access. Permission will not normally be granted in outline for new buildings in conservation areas.

If you decide an application is necessary it is advisable, though not essential, to instruct a professional to prepare the documentation. This will ensure that all procedures are undertaken properly, particularly the provision of ownership certificates and environmental assessments where appropriate, and that your case is made effectively taking into account all the relevant planning arguments. Without a full understanding of the issues that LPA must take into account when determining an application, you will not be in a position to maximize the chance of success.

In some rare cases the Secretary of State will **call in** an application to be determined by him rather than by the LPA. In this case there may be a public inquiry at the application stage. This will only be undertaken for major schemes likely to have a regional or national impact.

Planning permission may be granted subject to **conditions**. Conditions should only be imposed where they are necessary and reasonable and where they meet certain other criteria. All planning permissions will contain a condition stating that unless the development is begun within a specified time it will lapse. The period is normally five years, or in the case of outline permissions, two years after approval of the final reserved matter, as long as all the necessary reserved matter applications were submitted within three years of the permission.

Other conditions may include:

- conditions which take away permitted development rights;
- conditions prohibiting a change of use within the use class;
- conditions relating to access, landscaping and car parking;
- conditions on occupation, such as trading times, noise restrictions, prohibition of sale of certain goods.

Where **planning gain** is included there may need to be a **planning agreement** (sometimes known as a Section 106 agreement) detailing the works to be undertaken or payments to be made by the developer. It is important to take legal advice when negotiating the terms of such an agreement.

If planning permission is refused, granted subject to conditions, or if the application is not determined within eight weeks, you will have a right of **appeal** to the Secretary of State. Many LPAs take longer than eight weeks to determine applications and if you are sensitive to time you may wish to submit duplicate applications enabling an appeal to be made in respect of one, while the other is determined by the LPA. Appeals are often successful but this will be dependent on making the case effectively and the importance of instructing a professional is greater at this stage.

Only the original applicant can lodge an appeal and this must be done within six months of the decision. The appeal will normally be determined by an inspector, except for major developments where the Secretary of State may make the decision. Both the appellant and the local planning authority have the right to ask for a **local inquiry**, but the majority of appeals are handled by way of **written representations**. This is a relatively quick and inexpensive procedure and is usually the best route where the planning arguments and evidential issues are not complex. The appellant and LPA submit statements of case and observations within specified time limits on the basis of which the inspector determines the appeal. If a local inquiry procedure is chosen there are similar procedures for making written submissions followed by a hearing at which evidence is given and witnesses are cross examined. It is also possible to agree to deal with the appeal by a **hearing**, which is a less formal type of public inquiry.

4.9.3 Other permissions

In addition to planning permission, you will need to consider whether any of the following permissions are required.

(a) Listed buildings

The Secretary of State for the Environment holds a list of buildings of special architectural and historic interest. The buildings listed are classified into grades as follows.

Grade I   – buildings of exceptional interest;
Grade II   – buildings of special interest which warrant every effort being made to preserve them;
Grade II* – particularly important buildings in Grade II;
Grade III – this has no statutory force and is not used for buildings added since 1970.

Listed building consent is required for the demolition of a listed building or for its alteration or extension in any manner which would affect its character as a building of special architectural or historic interest. It is a common misconception that part of a building can be listed or that where part only is of merit this will be the only part affected by a listing. A listing always applies to the whole of a building even where special reference is made in the listing to particular features. The building will be looked at as a whole in determining an application for consent.

The benefits of the proposed development, including its architectural merit, will be weighed against other issues, including:

- the importance of the building, both intrinsically and relatively bearing in mind the number of other buildings of special architectural or historic interest in the neighbourhood and the contribution it makes to the local scene;
- the historical interest as well as architectural merit of the building;
- the condition of the building, the cost of repairing and maintaining it in relation to its importance, and whether it has already received or been promised grants from public funds;
- any alternative use for the site and in particular whether the use for some public purpose would enhance the environment and other listed buildings in the area.

It is a criminal offence to demolish, alter or extend a listed building without listed building consent.

A **building preservation notice** can be served by the local planning authority protecting buildings not listed for up to six months. If a listing is not confirmed within that time the protection elapses.

If listed building consent is refused, the applicant has a right of **appeal** to the Secretary of State for the Environment. The procedures are similar to those discussed above in relation to planning permissions.

(b) Conservation areas

Conservation area consent is required to demolish buildings in conservation areas, subject to certain exceptions, particularly small buildings. Other development does not require conservation area consent.

It is a criminal offence to demolish, alter or extend a listed building without consent.

In considering redevelopment in conservation areas the authority will have regard to the need to preserve and enhance the character and appearance of the area, and in particular:

- the contribution to the streetscape of the existing building and of the proposed building;
- the contribution of the building to the character of the area, such as historical associations and uses undertaken in the area.

If conservation area consent is refused, the applicant has a right of **appeal** to the Secretary of State for the Environment. The procedures are similar to those discussed above in relation to planning permissions.

(c) Other permissions

The following other consents are required.

- Consent to cut down, top, lop, uproot, damage or destroy trees subject to a **Tree Preservation Order**.
- Consent to display **advertisements**, which include a wide range of signs such as shop signs, estate agent's boards and hoardings. There is a wide range of permitted development rights which mean that planning permission does not have to be sought for many advertisements.

There are also special regulations governing the use of a site for caravans, storage of hazardous substances, and works to ancient monuments.

4.9.4 Enforcement

The LPA can use a range of powers to enforce planning control. Enforcement action cannot be taken however once certain time limits have elapsed, as follows:

- For operations (i.e. building works etc) – four years from the date the operations were substantially completed;
- For change of use of any building into a single dwellinghouse – four years following the change;
- For other changes of use or breach of a condition – ten years following the breach.

Where development has been undertaken without planning permission or in contravention of any conditions subject to which permission was granted, the LPA can use the following enforcement powers.

(a) Planning contravention notice

Where the LPA considers there may have been a breach of planning control, they may serve a planning contravention notice requiring information about activities occurring on the land. It is an offence to fail to comply with a planning contravention notice within 21 days of service and there is no right of appeal.

(b) Breach of condition notice

Where a planning permission has been granted subject to a condition which is not complied with the LPA may serve a breach of condition notice on the person responsible. The notice must specify the steps necessary to comply with the condition and the period for compliance. It is an offence to fail to comply with a breach of condition notice and there is no right of appeal.

(c) Enforcement notice

The LPA can serve an enforcement notice specifying the actions required to remedy the breach of planning control. It is an offence not to comply with an enforcement notice, and the LPA has the power to enter on the land to remedy the breach and can charge the costs to the owner.

There is a right of appeal against an enforcement notice. Lodging an appeal suspends the notice until the appeal has been decided. The appeal must be submitted to the Secretary of State before the date specified in the notice as the date on which the notice takes effect (at least 28 days after the notice is issued). Grounds of appeal include the following.

- There has not been a breach of planning control.
- Planning permission ought to be granted for the development or the condition ought to be discharged.
- The time limits on enforcement have elapsed.
- The requirements of the notice are excessive.
- The period specified in the notice is not reasonable.

(d) Stop notice

Where an enforcement notice has been served the LPA can, in some cases, also serve a stop notice preventing the continuation of any breach pending the taking effect of the enforcement notice and an appeal. This is designed to prevent an owner or occupier proceeding with development in breach of planning control while dragging out an appeal.

It is an offence to contravene a stop notice. There is no right of appeal but if the enforcement notice or stop notice is subsequently quashed on certain grounds or withdrawn, there is a right to compensation for losses incurred.

(e) Injunction

The LPA may apply for an injunction to restrain any actual or apprehended breach of planning control. This power is rarely used.

4.9.5 Compensation for planning decisions

Certain adverse planning decisions entitle the owner of the land to payment of compensation. The relevant decisions are limited and mainly comprise those where existing planning rights are withdrawn, such as where GPDO permitted development rights are withdrawn, where a planning permission is revoked or modified, or where a **Discontinuance Notice** is served limiting a use or requiring removal of buildings.

Similarly, in some cases where a planning decision effectively sterilises land and it is not capable of reasonably beneficial use in its existing state, an owner can force the local authority to purchase the interest by service of a **Purchase Notice**.

Please also refer to the section on compulsory purchase in Chapter 5.

4.9.6 Influencing planning decisions

Since the Planning and Compensation Act 1991, the development plan has become the primary consideration for judging development proposals. As a result, the importance of influencing the content of the development plan has become significant, and you may wish to ensure that your interests are represented in the process of producing the UDPs, Structure Plans and Local Plans.

First you need to find out what stage in the process the development plan is at. You should contact the LPA and county council and obtain copies of the relevant documents. If the plans have not yet been adopted there may be scope to comment on the policies included in them. If they have been adopted the process will start again in due course and you should find out the timing for the next consultation draft. It is normally advisable to take professional advice on the chances of changing any policies and to ensure that the arguments are made effectively.

The first step in influencing the process is normally to send a written representation. It may later be appropriate to attend an inquiry and in some cases to give evidence. In this case it is normally advisable to have professional representation.

In addition to influencing the formulation of policy, you may also be able to influence particular planning decisions at the time an application is considered. If a local owner or occupier applies for planning permission or other consents you can inform the LPA of your views and they will take these into account. If you are particularly concerned about a proposal, you may wish to lobby your local councillor, speak to councillors on the planning committee and encourage other people to lodge objections. In some cases, as a neighbour you will be informed of proposed development but usually you will have to rely on a notice posted on the land.

When trying to influence policy formulation or specific decisions is important that your written representations are based on planning arguments and not just personal opinion. Expressing a view that you do not like the design of a proposed building will be taken into account, but your comments will have a much greater impact if they are refer to recognized planning concerns. You will need to relate your comments to policies in the development plan and the other relevant planning arguments discussed above in Section 4.9.3.

4.9.7 Professional advisers

If you wish to know whether planning permission is required, need to obtain planning permission, or wish to affect planning policy and decisions, it may be necessary to take advice from a professional with appropriate expertise. Planning work is undertaken by, amongst others, specially qualified planners (members of the Royal Town Planning Institute), lawyers, surveyors (normally members of the Planning and Development division of the RICS) and architects. You should ensure that the adviser has the relevant qualifications and experience to deal with planning matters effectively.

4.10 BUILDING AND HEALTH AND SAFETY REGULATIONS

4.10.1 Building regulations

The Building Regulations 1991 as amended by the Building Regulations (Amendment) Regulations 1994 are relevant to most common building projects. The Regulations exist primarily to ensure the stability, durability and performance of building structures, fabric and services and to make provisions for the health and safety of persons within and without the building under normal and abnormal conditions.

The Regulations are deemed to be satisfied if the design and construction of the building comply with a series of Approved Documents published by the Department of the Environment and the Welsh Office. These documents address the following aspects of the structure, fabric and services of a building:

- structure
- fire safety
- site preparation and resistance to moisture
- toxic substances
- resistance to the passage of sound
- ventilation
- hygiene
- drainage and waste disposal
- heat producing appliances
- conservation of fuel and power
- access and facilities for disabled people
- glazing – materials and protection.

Relevant works may simply be the alteration of finishes (combustibility and spread of flame), alterations, refurbishment or change of use of an existing building, or the demolition and erection of a new building. It is advisable to consult an architect or chartered surveyor (building surveyor or quantity surveyor) to advise when and which statutory approvals are required and where necessary to act on your behalf in obtaining those approvals.

The enforcing agency is normally the local authority responsible for the site where the works are to be carried out and approval is granted by the Building Control Officer or City Engineers as applicable. It is a requirement that plans of relevant proposed works are deposited for approval with the local authority and that notices are served upon the local authority at various stages of construction to enable inspection to be carried out to ensure compliance with the Regulations.

Whilst the Building Regulations cover many aspects of construction, it should be noted that other authorities may also have control over construction activities. These may include the Health and Safety Executive, highway authorities and Public Health Inspectorate.

The fees charged by the local authority for Building Regulation approvals are calculated in relation to the total building cost of the works (excluding the designer's fees). For a project value of £100 000, for example, the fees will be in the order of £2700, payable in two stages as a planning fee before any construction works are commenced, and an inspecting fee on the commencement of works. In addition to Building Regulations approvals, most buildings require a fire certificate, confirming compliance with fire precautions regulations. In multi occupied buildings the landlord is normally responsible for obtaining this.

4.10.2 Health and safety regulations

Health and safety regulations encompass a wide range of issues including heating, ventilation, lighting, sanitary conveniences, drinking water and noise. In essence, it is the duty of every employer to safeguard, so far as is reasonably practicable, the health, safety and welfare at work of all his employees and there are specific regulations in this regard. Additionally, you must take reasonable care to ensure that visitors and trespassers are safe. In some cases, but by no means all, you may be able to discharge this duty by preventing trespassers gaining access or informing visitors of any danger.

The **Construction (Design and Management) Regulations 1994** concern the management of health and safety on construction projects including some maintenance works. They place duties on clients, project managers, designers and contractors to plan, coordinate and manage health and safety throughout all stages of a construction project to ensure that risks to health and safety, if not eliminated, are managed so that the minimum number of people are exposed to the lowest level of risk.

Prior to the start of a construction contract the client is obliged to appoint a **planning supervisor**, who can be either an in-house representative, project manager or other adviser. His first duty is to produce a **health and safety plan** to identify, as far as reasonably practicable, the risks that either exist on-site or are inherent to the work undertaken, so that resources may be allocated by the tendering contractors to mitigate these.

You will also need to appoint a **principal contractor**, who will normally be the main contractor. The principal contractor must build upon the health and safety plan, demonstrating how he will manage health and safety on site to eliminate or limit the risks. The principal contractor has a range of duties under the regulations, including ensuring that safe working practices are followed and that all personnel are properly trained to carry out their assigned duties. There are other requirements relating to monitoring subcontractors and preparation of construction records.

A surveyor, building services engineer or architect can advise on health and safety issues, and more general advice can be obtained from the Health and Safety Executive. Failure to comply with health and safety regulations may result in criminal liability.

4.11 ENVIRONMENTAL ISSUES

Occupiers can be affected by a wide range of environmental laws, depending on the use to which they are placing the relevant land and buildings, and the character of the area.

The purpose behind environmental laws is to control activities and potentially harmful substances so as to protect human health and the environment and to provide injured parties with a remedy when they have suffered damage. Environmental statutes cover the control, treatment and disposal of waste, the contamination of land, discharges to air, water and sewers, the control of hazardous substances, nuisance, noise and the protection of certain landscapes, flora and fauna. Breach of environmental law is generally a criminal offence. In serious cases, the maximum penalty is an unlimited fine and two years imprisonment.

As an occupier you will need to be sure that you hold, renew and comply with any necessary permits. For example, the discharge of any polluting matter into a stream requires a consent, as does the discharge of any trade effluent into a sewer.

You also need to ensure that you comply with all relevant statutory duties. In particular, there is a duty of care in respect of **waste**, which is explained in a statutory code of practice. This duty is in addition to any requirement to hold a waste management licence, and imposes various obligations on persons who produce, keep, store and transfer waste.

One environmental issue which can become complicated concerns **contaminated land**. The principal difficulty is the uncertainty which normally exists as to whether land is contaminated, whether it is resulting in any harm and establishing who caused it. Contamination can arise from natural processes (e.g. natural methane) or from human activities. Some land uses are associated with a high risk of contamination. Examples include gas works, coke works, landfill sites, metal mining and smelting, petrol stations and pulp and paper manufactures. However, the current use of the site is not the only factor: even on low risk sites, underground storage tanks may have leaked their contents into the surrounding soil. However, the current use of the site is not the only relevant factor: even though a site may not be put to a high risk use today, it may have been in the past, or contamination may have migrated from an adjoining site.

Contamination, may migrate from a site causing damage to a neighbouring occupier (for example, contamination of a borehole used to extract water for a brewery) thereby creating the grounds for a civil action. It may also amount to a criminal offence where, for example, surface water or ground water have been polluted. Under the Environment Act 1995, the local authority will have a more effective power to serve a remediation notice requiring contaminated land, and any affected water, to be cleaned up. If this notice is not complied with they will have the power to carry out the work themselves and recover the cost from the person who should have carried out the work.

Such a notice will be served on the person who caused or knowingly permitted the contamination to occur or remain. If, however, that person cannot be found, it can be served on the owner or the occupier of the land from which the contamination originated (although they cannot be made responsible for cleaning up land to which the contamination has escaped). It is important to note that, in some circumstances, choosing to ignore a problem which you know exists may make you liable in law as if you had created the problem in the first place. Accordingly, if you suspect your site is contaminated you should take legal advice or discuss the matter with the local authority Environmental Health department.

Remediation notices will only be served after the owner and occupier have been notified by the local authority that they consider it is contaminated and have discussed the matter with them. Clean-up works will only be required where significant harm is being or may be caused to human health or if pollution of the environment is likely.

These provisions will supersede a similar, but less powerful, legal regime which is not expected to be in place until early 1997. They have been greatly simplified here and if you receive a notice, or approach, from the local authority or other body concerning contamination of the land you occupy you should take legal advice. Those occupiers whose use of property involves waste or potentially contaminative uses should refer to one of the books listed in Further Reading for more detail on the subject, and advice from a specialist lawyer, surveyor or environmental consultant may be necessary.

4.12 NEIGHBOURING OWNERS

As an occupier either under a lease, or more commonly as an owner with a free-hold or long leasehold interest, you may encounter situations where you have to deal with neighbouring owners. Two important isssues are rights to light and party walls and are discussed below. It is important that you keep alert to development being undertaken around you in order to ensure that you protect yourself in respect of both your rights to light and party walls. In particular, this involves investigating any notices received or posted on neighbouring property. Equally, you should consider the rights of neighbouring owners prior to commencing redevelopment of your own premises. Bear in mind that if you are a tenant, you will probably be obliged to give copies of all notices received to your landlord.

4.12.1 Rights to light

In broad terms an easement of light can be acquired, either expressly, by implication or by prescription (acquisition over time), and in such cases an owner or occupier will be prohibited from obstructing the passage of light to neighbouring premises. This may affect the occupier:

- where you are undertaking alterations or redevelopment and need to ensure you do not obstruct an neighbouring owners' right to light;
- where an adjoining owner undertakes works which interfere with your right to light.

Infringement of a right to light could result in an injunction stopping the work or a damages claim. If you are undertaking major works or an adjoining owner is doing so and you are concerned that any rights to light may be prejudiced, you should take the advice of a lawyer or specialist building surveyor. Rights to light are potentially a valuable asset which may need to be protected.

4.12.2 Party walls

If you or an adjoining owner or occupier are undertaking construction works which will affect a boundary wall used by both adjoining properties you may have to deal with a party wall dispute. You will normally have a right of support from the wall and may have a duty to repair it, or to share in the costs of repairs. In London there is a detailed statutory framework governing party wall issues, and elsewhere the common law provides a number of similar rights to property owners. New legislation has been introduced to extend the arrangements in London to the rest of England and Wales, but this has not yet been implemented.

Under the London Building Acts, extensive rights are given to a building owner to carry out works to existing party structures subject to the service of notices on the adjoining owner. The rights include repairing, underpinning, demolishing and rebuilding. The developer has to make good any damage caused to your premises and must not cause any unnecessary inconvenience. If a notice is served on you as an adjoining owner, you have certain rights to serve counter notices requiring amendments to the specification of works.

Where an owner wishes to build a wall on a boundary of his land previously not built on, there are regulations relating to consent and apportioning the subsequent cost of repairs.

If you receive a notice relating to a party wall structure you should take the advice of a lawyer and/or building surveyor. You may need to serve a counter notice or take other proceedings to protect your position. If an adjoining owner undertakes works to a party wall without consulting you, you should check with a lawyer or building surveyor whether this is in contravention of the rules.

Where an adjoining owner does not give consent to works within 14 days following service of a notice, there is a procedure for resolving disputes. If you are undertaking works which will affect a party wall, you should take the advice of a lawyer and/or surveyor to ensure you comply properly with your obligations.

As a rule, matters are usually mutually agreed between party wall surveyors acting on behalf of each building owner and the agreement is recorded in a party wall award.

4.13 ENFORCING COVENANTS

If faced with a breach of any of the covenants in the lease, there are a number of remedies to which the landlord or tenant can turn. Leases afford additional protection over other business contracts; there are special common law and statutory remedies as well as other special rights usually included expressly in lease documents.

4.13.1 Landlord's remedies

The remedies available to landlords offer them considerable power to enforce tenants' covenants, and are significantly in excess of those available to tenants. The range of landlord's remedies is outlined in Figure 4.1.

The landlord's first step in dealing with a breach of covenant will normally be to inform you of the breach and ask you to remedy it. If you do not, he will send a **letter before action**, sometimes known as a seven-day letter, formally giving you seven days in which to remedy the breach. If you do not do so the landlord can choose to use any of the remedies discussed below. The remedy adopted will depend on the nature of the breach and the other circumstances.

The general rule for leases is that the **limitation period** for instituting proceedings is 12 years; failure to sue within this period will render the claim statute-barred. However, there is a special rule in relation to rent and other sums reserved as rent under the lease (usually service charge, insurance and some other payments). For these payments, the period in which the landlord must

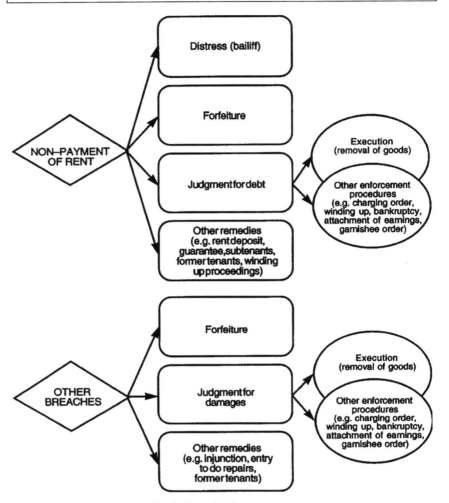

Figure 4.1 Landlords' remedies for breach of covenant.

issue proceedings is six years from the date the payment became due, unless the debt has been acknowledged at a later date, in which case the six-year period begins to run again from the date of acknowledgement.

(a) Distress

Distress is an ancient remedy which entitles the landlord to enter the premises where rent is unpaid and take possession of the tenant's goods to satisfy the amount of the outstanding rent. The landlord will normally employ a **certificated bailiff** to carry out the distraint.

The bailiff can remove goods found in the demised premises and impound them either on or off the premises. Most commonly the goods will be held on the premises, a procedure known as **walking possession**. In this case it is a criminal offence for anyone other than the landlord (or someone acting on his behalf) to remove them during the period specified in the notice of distress (a minimum of five days). In either case the landlord is entitled to sell the goods after the specified period has elapsed, unless during this time you have paid the outstanding rent.

The restrictions governing the use of distress and the procedures the bailiff can adopt are complex and legal advice should be taken. The following are some general points to keep in mind.

● If the landlord levies distress wrongfully you may be able to take an action to recover the goods (a procedure known as 'replevin') or to obtain an injunction preventing further distress. You may be entitled to damages where the

The landlord usually has a right to levy distress.

landlord wrongfully distrains and sells the goods and you may be entitled to double value in this situation. In some cases, you may be entitled to much larger damages where wrongful distraint has caused business loss.

- The landlord cannot levy distress where the outstanding sums are genuinely in dispute. This is most common for non-rental payments reserved as rent in the lease, for example where the tenant disputes a service charge calculation. In general, it is safest where items are in dispute to make payment 'without prejudice' and obtain a refund once the matter is agreed, although in some cases this may not be advisable (especially if you have reason to believe the landlord may become insolvent) or may not be necessary (if you have a right of set off under the lease). The landlord cannot distrain for rent which is lawfully withheld in accordance with a right of set off (a right to withhold rent where the landlord is in breach of covenant) – see tenant's remedies under Section 4.13.2 below.

- In some cases distress can be used in respect of other sums payable under the lease, such as service charge and insurance, if these are reserved as rent. Where such sums are not specified as a definite figure in the lease there may be scope to oppose the distraint and legal advice should be taken.

- A **third party** can make a claim for goods owned by them and may be able to recover possession or pursue a damages claim. The landlord is entitled to remove goods if they are in the reputed ownership of the tenant, and the onus is on the tenant or third party to notify the landlord that the goods are in the third party's ownership.

- Certain items are **privileged** against distress. These include tenant's trade fixtures, loose money (but not money stored in a safe) and wild animals! Any item actually in use at the time cannot be distrained against.

- Be careful about taking any precipitous action to avoid the bailiff removing goods. The landlord may be entitled to up to treble the value of the goods where the tenant removes them with the intention of avoiding distraint (known as '**rescue**'), and where the tenant takes back goods that have been seized or impounded (known as '**pound breach**') and there may also be criminal penalties. While the landlord can only normally levy distress on goods located in the demised premises, the exception is where you remove the goods clandestinely and fraudulently in order to avoid distraint; in this case the landlord can enter onto the premises to which the goods have been moved.

(b) Forfeiture

Almost all modern leases contain a proviso allowing the a landlord to re-enter in the event of a breach of covenant. Where there is such a clause, the landlord has a right to forfeit the lease under the Law of Property Act 1925. This is subject to a tenant's right to relief from forfeiture in certain circumstances (see below).

The right of the landlord to forfeit the lease is not enforceable until he has served a notice under section 146 of the Law of Property Act 1925 giving the tenant notice of the breach and a period in which to remedy it, except where the breach is non-payment of rent, in which case the Section 146 notice is not required. Once the period has elapsed the landlord can make a claim for possession either by bringing an **action for possession** in the courts, or by taking **peaceable re-entry**. The latter procedure simply requires the landlord to enter onto the premises and take over possession by changing the locks. The landlord cannot do so if there is someone on the premises at the time challenging re-entry.

Once the landlord applies for or takes possession, the tenant or any subtenant can apply for **relief from forfeiture**. The courts tend to favour tenants, especially where the forfeiture would deprive them of a valuable asset. However, the tenant must show a willingness and ability to comply with their covenants and persistent material breaches will not normally be tolerated. Where the breach is non-payment of rent, payment of all or a significant part of the arrears will often be a condition of the relief. Where a subtenant is granted relief, they will normally have to take on the terms of the headlease (save for the term, which will be that of the sublease) and may have to pay the headlessee's rent arrears even if they themselves are not in arrear.

Forfeiture has not been widely used in recent years because of the decline in the property market. Where the landlord is unlikely to be able to relet the premises quickly or at a similar rent, they will rarely resort to this remedy.

There are some situations in which the landlord loses his right to forfeit. Legal advice should be taken but broadly the principles are as follows.

- The landlord may have waived the breach of covenant. A **waiver** occurs if the landlord, with the knowledge of the breach, acts in such a way as to recognize unequivocally the continued existence of the tenancy. Examples of acts of waiver are acceptance of rent for any period after the breach, communication between the landlord and tenant in relation to a consent to alienation or alteration or enforcement by the landlord of other covenants. Knowledge by the landlord's agent is held to be knowledge by the landlord, and generally, an act by the agent, such as the demand or acceptance of rent, will waive the breach. Where rent is paid by standing order, the automatic receipt of rent into the landlord's account may not be a waiver, provided the rent is returned as soon as the landlord becomes aware that it has been received.

  Waiver only applies in respect of once-and-for-all breaches such as carrying out alterations and assignment or subletting without consent, non-payment of rent for a particular period, and insolvency. Continuing breaches include failure to repair or insure, and breach of the user covenant. These breaches continue until they are remedied and the landlord will not lose the right to forfeit by an act of waiver.

- The landlord may not have complied with all technical requirements. An example is a requirement formally to demand rent before possession can be taken. This requirement is excluded in most modern leases, but where it is not and the landlord omits to make a formal demand he may lose the right to forfeit.
- The tenant can prevent the landlord from peaceably taking possession by appointing security guards or other staff to stay on the premises and prevent re-entry. Re-entry is not legal where the landlord meets such resistance and he will have obtain a court order if he wishes to secure possession.
- The tenant may be able to obtain an **injunction** preventing peaceable re-entry.

(c) Damages

The third remedy is an action for breach of contract. This is known as an action in damages as the landlord will make a claim for damages to compensate him for the breach. The damages usually comprise the loss directly associated with the breach together with costs and interest at a statutory rate. For breaches other than non-payment of rent damages are calculated by reference to the cost of remedying the breach. In the case of a breach of a repairing covenant the damages cannot exceed the diminution in the value of the landlord's interest resulting from the breach. Therefore, where the landlord is expected to redevelop the premises and will suffer no loss as a result of the breach of a repairing covenant, no damages are payable (Section 5.3)

In making the claim the landlord has to undertake the following steps.

- Section 146 notice (for breach of the repairing covenant only). Where the breach is of a repairing obligation, the landlord must serve a Section 146 notice specifying the breach, which will normally require a schedule of dilapidations (Section 5.3). Where the provisions of the Leasehold Property (Repairs) Act 1938 apply (Section 4.5.1 above) the notice must also specify that the tenant can claim the protection of this Act by service of a counternotice. Where you claim the protection of the Act the landlord may be unable to proceed with a damages claim.

  You will be entitled to a reasonable period in which to remedy the breach. The period will depend on the nature of the lack of repair, but except in the case of minor repairs will usually be several weeks.

  A Section 146 notice is not required where damages are being claimed in respect of a breach other than lack of repair. For other breaches the first step is normally a 'letter before action' from the landlord's solicitors outlining the claim as a prelude to legal action. The landlord does not usually have to send such a letter and may move straight to proceedings.
- Proceedings. If the tenant fails to comply with the Section 146 notice or letter before action, the landlord can serve proceedings. You will have a limited

period in which to protect your interests and should consult a lawyer immediately.

The landlord may be able to obtain summary judgment after a short hearing if it is clear that the tenant does not have an arguable defence to the claim. Otherwise there will be a number of procedural stages leading ultimately to a full trial.

- Enforcement. If the landlord is successful in obtaining judgment against the tenant in the County Court this will be recorded as a County Court Judgment and will affect the tenant's credit status in the future. (There is no comparable record of High Court judgments.) The landlord still has to enforce the judgment, for which he has the following options.
  - Execution on goods – removal of the tenant's goods to the value of the judgment, wherever the goods may be (i.e. not just at the demised premises as for distress described above). This is carried out by officers of the court.
  - Garnishee order – this compels payment of cash from a bank account, or other debtors.
  - Charging order – this charges a property owned by the tenant and payment of the debt must be made from the proceeds of any sale of the property (unless there are any prior charges).
  - Attachment of earnings order – this is an unusual method and compels an employer of the tenant to make deductions from their earnings and forward them to the landlord towards satisfaction of the debt.
  - Winding up or bankruptcy proceedings.

(d) Other remedies

There are a number of other remedies available to a landlord in some cases. The landlord may have the benefit of a **rent deposit**. Landlords are often reluctant to use these as a first resort because of the difficulty in making the tenant replenish them. Where there is a bank **guarantee** or guarantee by another party, such as a parent company, individual director, or former tenant the landlord may pursue these parties in addition to or instead of proceeding against the tenant.

Where a former tenant has a liability under the rule of **privity of contract**, which applies to all leases granted prior to 1 January 1996, the landlord may choose to pursue these tenants.

If the tenant has granted a sublease the landlord can serve a notice on the subtenant under the Law of Distress (Amendment) Act 1908 requiring payment by the subtenant of the sublease rent directly to the landlord to satisfy the amount of the tenant's arrears.

Where the breach is other than non-payment of rent, the landlord may have further remedies available to him. It is common in modern leases for a landlord to have an express **right to enter** onto the premises to remedy certain breaches, particularly to undertake repairs. The landlord will normally be able to charge the costs involved in such action to the tenant.

In some cases he may be able to obtain an **injunction** to prevent you from doing something prohibited in the lease, such as undertaking alterations, committing a nuisance or to require you to do something to remedy the breach. He may in some situations be able to obtain an order for **specific performance** obliging you to comply with a positive covenant.

The landlord can apply to the court for a tenant company to be **wound up** without first obtaining judgment, if the amount of the debt is definite and is greater than £750. Service of a **statutory demand**, is the first step in the process.

4.13.2 Tenant's remedies

The remedies available to a tenant are more limited than those available to a landlord, mainly because they exclude the most powerful remedies, distress and forfeiture. Landlord's breach of covenant is not a common problem as most landlords wish to maintain the investment value of their properties by repairing and maintaining common areas and providing other services properly. Problems arise mainly where there are significant service charge shortfalls which can discourage the landlord from providing the proper level of service, or where the landlord becomes financially insecure or insolvent.

In all cases when deciding whether to take action to enforce a landlord's covenants you will need to consider the following points.

- What is the amount of your loss? This will limit the amount you will receive in a damages action against the landlord, guarantor or previous landlord. Where your financial loss is small an action in damages may not be worthwhile. The amount of the loss will be considered by the court in determining whether to grant specific performance or an injunction, although in some cases other factors will be more important.
- What are the financial resources of the parties? Where you are making a claim for damages you will need to consider which party or parties, if any, should be pursued. If the landlord is financially insecure or insolvent it is probably not worth suing him and you would choose to pursue a guarantor or original landlord in preference.

Bear in mind that the remedies that can be used in response to a breach of covenant are limited in the case of insolvency. If you wish to enforce your rights against a landlord who is in Receivership, Administration, Administrative Receivership or Liquidation, some of the above remedies may not be available. Equally, if you become insolvent the landlord may not be able to use certain types of proceedings against you. Legal advice should be taken.

(a) Set off

The most powerful remedy available to a tenant is the right to set off the cost of remedying the landlord's breach against sums owing to the landlord. This may be a right expressly reserved in the lease or may be available at common law. Most modern commercial leases expressly exclude the right. Where there is no express right, the circumstances in which set off can be used are limited and it is essential that you obtain legal advice before using this remedy.

(b) Damages

The tenant can sue for damages but tenants often find difficulty in demonstrating that the landlord's breach has resulted in a financial loss. Summary procedures will normally be available and are usually relatively quick and inexpensive.

(c) Specific performance and injunction

These remedies are more widely used by tenants than by landlords, mainly because they are appropriate for enforcing non-financial covenants, such as the covenant to repair, to provide services or to give quiet enjoyment, where financial compensation is not a satisfactory remedy. An order for specific performance requires a positive covenant to be carried out and can be used in the case of continuing breaches. Injunctions are either mandatory (positive) or prohibitory (negative) and order the landlord to do, or refrain from doing, a particular thing.

(d) Other remedies

In some situations the tenant may be able to make use of other remedies. There may be a **guarantee** of the landlord's covenants by a third party. If the current landlord is not the original landlord on the lease you may be able to pursue an **original landlord** under privity of contract, unless they have been released under the provisions of the Landlord and Tenant (Covenants) Act 1995 or the landlord's covenant is expressed to be binding only for so long as he holds the reversion.

In some unusual cases the landlord may have placed funds in an **escrow deposit** which can be accessed in the event of a breach of covenant.

4.14 MANAGING COSTS OF SURPLUS PROPERTIES

The problem of surplus property has become a significant issue for many occupiers in recent years. Downsizing and restructuring have resulted in large amounts of space surplus to requirements.

The problem of surplus space needs to be addressed proactively and methods of achieving disposals are discussed in Chapter 5. However, while the surplus property remains as a liability the occupier needs to find ways of mitigating the costs. The following are some strategies that should be considered.

- Confining your space usage to a particular part of the building can reduce outgoings. In particular, the rates liability can often be reduced by obtaining empty rates on the vacant areas. Additionally, utilities, maintenance and services costs may be reduced. Where you are able to separate areas, you may be able to achieve sublettings of these parts (see Chapter 5).
- Consider ways of reducing the rates liability. Ensure that you notify the rating authority as soon as you vacate space. If only part of the premises are vacant ensure that all the vacant areas are located together to ensure that you obtain empty rates. It may be appropriate to remove structural items or services to eliminate the rates liability ('constructive vandalism'). You will need to determine whether this will contravene your lease obligations but if it does you may be able to obtain the landlord's agreement on condition that you remedy all damage subsequently or compensate him. You will also need to ensure you comply with planning, building and health and safety regulations. (Sections 4.9 and 4.10 above).
- Discuss with the landlord ways of reducing the service charge. Identify any unnecessary services or any that could be provided to a lower standard. In a multi-occupied building other tenants may oppose such changes unless they too have cost concerns (Section 4.3.2 above).
- Ensure that any services you run yourself have been reduced to an appropriate level to take account of your having vacated. Where appropriate, terminate water and energy supplies and disconnect air conditioning, heating and other systems to minimize costs. You may be able to terminate service contracts on plant and equipment and move onto a less frequent servicing regime.
- Consider the level of insurance cover to ensure that the contents sum insured is not excessive. Additionally, if you are paying a higher premium as a result of the premises being unoccupied obtain competitive quotations to see if other policies would be cheaper and consider whether it would be economic to use some form of security service (Section 4.4 above).
- If you are a company with limited assets, a difficult question arises as to which creditors (such as the landlord and the rating authority), if any, ought to be paid. This may be a matter of particular concern to the directors of the company who may find themselves liable for the company's debts. Companies in financial difficulty may find it appropriate to consider voluntary liquidation and disclaimer of the lease.

FURTHER READING

Aldridge, T. M. (1992) *Boundaries, Walls and Fences*, Longman Law, Tax & Finance, London.

Anstey, J. (1988) *Rights of Light*, Surveyors Publications, London.

Barrett, P (1996) *Facilities Management: Towards best practice*, Blackwell Science

Beaumont, B. (1993) *Arbitration and Rent Review*, The Estates Gazette, London.

Bernard Williams Associates (1995) *Facilities Economics*, Building Economics Bureau Limited, Bromley, Kent.

Brand, C. (1994) *Longman Practice Notes: Planning Law*, Longman, London.

British Council of Shopping Centres (19XX) *Service Charges in Commercial Properties: A Guide to Good Practice*.

Brunt, P. D. (1988) *How to Cut Your Business Costs*, Kogan Page, London.

Department of the Environment/Welsh Office (1991) *Environmental Protection Act 1990: Waste Management: the Duty of Care: a Code of Practice*, HMSO, London.

Drivers Jonas (19XX) *Whose Money is it Anyway?*, Drivers Jonas, London.

Jones Lang Wootton (annual) *OSCAR (Office Service Charge Analysis)*, Jones Lang Wootton, London.

Male, J. M. (1995) *Landlord and Tenant*, Pitman Publishing, London.

McManus, F. (1994) *Environmental Health Law,* Blackstone Press, London.

Park, A. (1994) *Facilities Management*, The Macmillan Press.

Smith, P. F. (1993) *The Law of Landlord and Tenant*, Butterworths.

<table>
<tr><td>**5**</td><td>**Exit and lease renewal**</td></tr>
</table>

KEY POINTS

- Disposal strategies and effective marketing of surplus property
- Dealing effectively with expiry and renewal of protected and unprotected tenancies
- Legal and practical issues when vacating

In Chapters 1 and 2 the process of developing property strategy, assessing needs and objectives and planning space usage are discussed. This planning activity has to continue throughout the period of occupation with a continual reassessment of short- and long-term needs. This chapter focuses on the end of the occupational process, when contractual obligations come to an end, or when there is a need to exit from the premises.

Disposals involve much the same principal stages as acquisitions. The first step is to determine needs and objectives and the options available for meeting these. Refer to Chapters 1 and 2 for an explanation of the process of evaluating options using qualitative and quantitative analysis to establish a preferred strategy. Where you have the option to stay in the premises, you normally need to start with a stay-versus-move analysis, before moving on to analysing the relocation options.

The following sections provide the practical and legal framework for dealing with exit and renewal issues, including disposal strategies, lease expiry, dilapidations and reinstatement, and compulsory purchase.

5.1 SURPLUS SPACE STRATEGIES

In general, there is a conflict between business plans, which tend to relate to the short to medium term, and property commitments which are long term and inflexible. Surplus space has become one of the most significant property problems for occupiers in recent years. In the 1980s business growth was accompanied by expansion of corporate property portfolios. The decline in economic activity combined with ongoing changes in working practices have led to a large pool of surplus space representing a substantial drain on business.

The first step in addressing the problem is to go back to the acquisition process. An unstructured approach at acquisition is conducive to short termism. By adopting the structured approach described in Chapter 2, business objectives are identified and prioritized, alternative methods of meeting them are analysed, and the chosen option is implemented effectively. While the extent to which business objectives can be met depends on the properties and transactions available in the market, the structured acquisition approach ensures that long-term objectives are taken into account and where long- and short-term needs conflict an appropriate compromise is achieved.

The problem of surplus property is likely to be endemic for the foreseeable future as a result of a long-term decline in demand due to technological development and changing working practices. Occupiers need to assemble an armoury of weapons for addressing the problem.

Instructing a commercial agent to distribute particulars and advertise in local publications is normally a relevant part of the process, but if the problem is to be addressed effectively more innovative solutions need to be considered. In many cases such techniques require an investment of capital, which often becomes a stumbling block. However, when the size of the problem, and the consequences for corporate profitability are considered, capital investment is often worthwhile. One of the main obstacles to such investment is that the costs of vacant properties do not always have to be recognized in company accounts as a one-off charge if there is an expectation of a disposal, and can instead be expensed annually in these circumstances. In the short term a large investment can have a serious negative effect on the profit and loss account, and this can sometimes outweigh the benefits of an investment which makes sense in purely economic terms. A change in the way in which lease payments are shown in company accounts, so that the full obligation is capitalised and shown as a liability, has been mooted, but is unlikely to be implemented for some time. In large companies, short termism can be avoided by allocating a provision to business units which they can use to fund disposals.

5.1.1 Disposal strategies

A disposal strategy needs to be formulated taking into account the particular attributes of the property. It is important to think laterally to take advantage of all opportunities.

The disposal options available depend on whether you hold a freehold or leasehold interest, as shown in Table 5.1.

The disposal route adopted and marketing methods used will depend on the objectives of the disposal and the legal interest held. Where you hold a lease, this may have value, for example if the rent is below the market level; alterna-

Table 5.1 Disposal options

Interest	Disposal method	Description
Freehold	Sale	Transfer of ownership
	Sale and leaseback	Transfer of ownership and simultaneous grant of an occupational lease back to the occupier
Leasehold	Surrender	Termination of all liabilities under the lease
	Assignment	Transfer of obligations and rights under the lease to new occupier
	Subletting (whole or part)	Remain tenant under headlease but transfer occupation on agreed terms to subtenant

tively you may need to pay to discharge your obligations where the lease is over-rented or there are other onerous obligations.

In all cases it is essential to prepare the ground effectively for the disposal. This means assembling all relevant information, including details of the legal interest to be marketed and documentation, all cost information and technical information relating to construction, services, planning, statutory approvals, landlord and tenant matters and so on. This information will be necessary for you when assessing your liabilities, and will be important when you come to preparation of marketing details and negotiations with prospective occupiers/owners.

In some cases you may need to undertake more extensive work prior to marketing. Where redevelopment or change of use is required to improve the disposal prospects, you may need to obtain planning permission, buy in other interests and deal with legal issues. Similarly, where environmental problems will be an issue, commissioning an independent environmental audit may assist marketing. In all cases you need to adopt a structured approach to ensure you maximize the speed and proceeds of disposal.

The specific strategies to be adopted will depend on the nature of the property and your interest, but the following are normally the main ones to consider.

(a) Lease surrender

The first step in achieving a disposal of a lease should normally be to speak to the landlord about a surrender. This will establish whether a surrender is negotiable and the likely price, which can be used as the parameters for evaluating other types of disposal transactions. A surrender is often the preferred disposal route for an occupier because it terminates all liabilities without the problem of contingent liabilities that can be associated with assignments and sublettings.

In most cases a landlord will accept a surrender of the lease if the price is right but in some cases the premium the landlord requires is uneconomic. The landlord may require the whole of the present value of the lease commitment to be paid in a reverse premium. If a surrender is in accordance with business strategy it may still be worthwhile on these terms but you should normally be able to negotiate an allowance to reflect the landlord's possibility of reletting. The landlord's assumptions are usually relatively cautious, with long estimated marketing periods and low rental levels for a new letting. In theory it should be cheaper to pay off the landlord than market the premises yourself as the landlord should be more skilled and experienced in achieving the best terms in the shortest possible time; in practice, the landlord may require a high level of compensation for the additional risk he will be taking. You should aim to negotiate a transaction reflecting realistic assumptions in respect of void periods, rental levels and costs.

Financial analysis can be a helpful tool in persuading the landlord to accept a surrender at a favourable premium. Not only will it enable you to understand fully the landlord's perspective, but by taking control of the analysis you can sometimes use the figures in your favour to show why the landlord should accept a particular figure. Figures can be sensitive to small changes in assumptions and by preparing the analysis on appropriate assumptions the figures can help to support your case.

It is important to establish clearly the options open to the landlord. In some cases he may have little scope to enforce your covenants, for example where the lease is held by a subsidiary holding company with few assets. Similarly, he may be better able to pursue a previous tenant or guarantor where they have a continuing liability. The chance of the landlord enforcing your obligations should be taken into account in deciding how much you are prepared to pay for a surrender. If the cost of proceedings would be disproportionate to the size of your obligation, for example where the legal issues are not straightforward, this will affect the landlord's negotiating position. It is advisable to take legal advice to confirm your obligations and the landlord's scope to enforce them.

(b) Maximize value and marketability

In some cases a change of use, redevelopment or refurbishment can make a significant impact on marketability and the rental level or sale price that can be achieved. Where you hold a lease that does not have any value, often because it is over-rented or has a short period to expiry, you may be able to find ways to improve the marketability. Equally, where your interest does have a market value, you will wish to maximize this. Estate agents will commonly advise on minor refurbishment possibilities, but may not always identify more substantial redevelopment prospects, change of use potential and other strategies to improve value.

You should consider both **physical issues** and **tenure issues** which affect value. Physical issues include redevelopment and change of use. Tenure issues are those relating to the legal interest you hold. Where the period left to run on the lease is short you may need to discuss with the landlord the possibility of an extension so that this can be offered to potential occupiers. If the lease is protected by the Landlord and Tenant Act 1954 and the landlord does not expect to oppose the grant of a new lease on expiry under one of the permitted grounds (see below), the landlord may be prepared to agree an extension now. If there are any particular terms that may prejudice marketability it is advisable to discuss varying the lease with the landlord, for which a payment may be required.

In order to maximize the marketing potential of vacant property it is important to consider the underlying demand and supply relationships in the local property market and for this you may need to enlist the help of a market analyst. **Market analysis** services are provided by specialist economic consultants and the large surveying practices. By rigorously analysing long-term and short-term demand trends you will be in a position to maximize the value of the site or property. Analysis of the structure of demand can also be used as the first stage in a targeted marketing campaign. By identifying growth sectors and potential movers a target list can be developed which can be used as the basis for marketing.

(c) Targeted marketing campaign

In most cases there will be a need to implement targeted marketing using estate agents. The agents bring their knowledge of the local market and act as intermediaries bringing together buyers and sellers.

However, it is not enough just to appoint an agent and leave them to get on with the job. To maximize the success of the project, it is important to adopt a targeted approach, identifying potential purchasers from the agent's databases, lists of local occupiers and any research specially commissioned from economic analysts who can identify potential movers (see market analysis above), and making use of your own contacts and knowledge of your own and other industries. While there is a role for less focused promotion, such as distribution of particulars and advertising, this alone will not achieve a successful disposal in the shortest possible time or on the most favourable terms. A commercial property agent can advise on the **marketing strategy** for a particular property and the terms on which the property should be offered, but you should encourage them to identify innovative and proactive solutions.

An essential part of the marketing strategy is determining the terms to be offered. In some cases, this is limited to price, but in many cases, particularly for disposal of a lease by assignment or subletting, other terms will need to be considered. Ideally you will wish to offer attractive competitive terms without underselling; in practice it can be hard to achieve this balance. Having a clear idea about your priorities and explaining them properly to your advisers will play an important role. For example, if a quick disposal is your primary goal, you may not wish to waste time testing the market by quoting unattractive terms. Market analysis and the agent's advice will assist in achieving the right balance.

This is not the forum to discuss the detail of the marketing process but the central issues of relevance to an occupier are discussed in the following sections. For more detailed information on the technicalities of marketing refer to a specialist textbook.

(d) Bulk disposal

Another marketing technique that can be appropriate where an occupier holds a number of surplus properties is a bulk disposal route. Even in relatively poor market conditions there are usually acquisitive occupiers, looking to expand their property portfolios quickly, most commonly in the retail sector but sometimes in others also. Property companies are sometimes prepared to take on a portfolio in its entirety, normally where some of the properties are income producing or have good letting prospects. This would normally have to be a trading company looking for high risk/return secondary investments. A purchaser can be found with the help of market analysts (described above) and commercial agents, but it is often appropriate to make your own enquiries and to use your own business contacts.

The advantages of the bulk disposal include:

- a portfolio-wide disposal achieved quickly reducing the management time involved in the disposal process;
- disposal of less attractive properties can sometimes be achieved on favourable terms when they are grouped with more marketable ones.

The key is identifying a company whose property needs coincide broadly with the properties in the portfolio.

5.1.2 Disposal methods

The most common method of sale is the private treaty route and in most situations this is the most appropriate. Where a lease is being sold, the private treaty route is normally the only appropriate one, but in some rare situations a tender can be used. All methods can be used in the case of a sale of a freehold or long leasehold interest. Your agent, property adviser or lawyer can advise you on the most suitable approach in the circumstances.

- Private treaty. This is the usual form of sale transaction. The purchaser makes an offer subject to contract which, if accepted, is translated by a lawyer into a documented transaction. The system is flexible and allows for negotiation but the risk of the transaction falling through continues up to completion which can result in wasted time and cost.

  The vendor may be asked to sign a **lock-out agreement** during the lead up to exchange of contracts, preventing negotiations with other parties for a limited period.

Bulk disposals.

- Auction. This is most suitable where a quick sale is required or where there is a need to demonstrate a competitive bidding situation, particularly where fiduciary interests are involved, for example for sales of repossessed properties, or those by trustees or executors.

  The advantage is that a sale is usually effected immediately and there is no delay for completion of a contract. The main problem with this method is that you cannot be sure to obtain the best market price. The bidders at an auction are limited to those who can complete a sale immediately and who have been able to invest time and money in investigating the property prior to the auction. With such a limited market the price obtained can often be depressed. If the highest bid is above any **reserve price** (the minimum sale price you specify), you may be bound to complete the transaction. There can also be significant time and cost involved in providing prospective purchasers with enough information for them to be able to bid.

● Tender. Offering the property for sale by tender can be used to achieve a quick disposal, particularly where you have identified a number of bidders for a property and wish to see their best offers. You set a deadline by which they have to submit written bids.

This method has some of the advantages of the auction in achieving a quick transaction, without the obligation to proceed to a transaction on the terms offered. Purchasers are motivated to submit the highest possible bid as they are not aware of the strength of competition. You are able to assess the different bids together. However, the method is not popular with purchasers because of the need to make investigations before bidding and because of the weakness of their position, bidding 'blind'. As a result the price achieved may be depressed.

It usual for a tender to be followed by a private treaty transaction on the terms submitted in the tender. In some cases a **formal tender** may be used where all the terms offered by the purchasers are quoted in the tenders and a binding contract is effected if these terms are accepted. This formal tender route is rigid and rarely used in practice.

5.1.3 Instructing agents

(a) Choosing an agent

Personal recommendation is usually the best basis on which to choose an agent. In the absence of this, it is usually advisable to use an agent with a strong local profile, although local presence is only one of a range of important skills, in particular professionalism, general market understanding and negotiating skills. You will need to weigh up the focused attention and specialist skills that you can get from some small niche firms, against the regional or national coverage and resources that a large firm can offer. Where it is important to combine local and national exposure, a multiple agency, joint agency or subagency instruction may be appropriate (see below).

Your agent's role should not only be to find a new occupier or owner but also to advise you on the terms of the transaction and normally to undertake negotiations. You will need confidence in his expertise and professionalism as a marketeer, negotiator and adviser.

(b) Types of instruction

When deciding on the type of agency instruction to adopt you need to consider the objectives of the project, and the best agency structure to achieve these. In particular, you need to consider the extent to which you wish the agent to a take an advisory role and to act for you in negotiations, the need for exposure contrasted with the need to avoid over exposure (a property can seem unattractive if it is seen to be aggressively marketed), and the importance of motivating the agent with the fee arrangement. The following are the main options.

- Sole agency. You instruct only one agent who is paid a fee if he is the effective cause of the disposal. The arrangement is conducive to a professional advisory relationship between agent and principal, and reduces the potential for commission disputes that can arise with other agency structures. Additionally, the agent is likely to be motivated to work hard for a sale as he will not be concerned about being 'pipped at the post' by another agent. The main problem with this structure is the potential lack of exposure that can result from only one agent promoting the property.

   A variation on the sole agency arrangement is **sole selling rights**. In this type of instruction the agent receives a commission even if he is not the effective cause of the sale. This is normally only appropriate where the agent will be involved in significant negotiation and advisory work even if he does not introduce the purchaser.
- Multiple agency. In this case you instruct more than one agent with payment of a fee only to the agent who is the effective cause of the disposal. The main advantage is the greater exposure that can be achieved. This must be weighed against the risk of over exposure and the potential of the agent not investing as much effort in the property where there is a risk he may not receive a commission. The principal/agent relationship can be compromised with this arrangement as you will be dealing with a number of agents and there is the risk that none will see themselves as having a professional duty to you in an advisory role. The fee is usually higher to reflect the risk of abortive work.
- Joint sole agency. This structure overcomes some of the problems of the sole and multiple agency structures, but has the disadvantage of additional cost. A number of agents can be instructed and all are paid a commission if a sale is effected regardless of who introduces the purchaser. A high level of exposure can be achieved and the scope for commission disputes and lack of motivation that can hamper a multiple agency instruction is avoided. The arrangement is often used where there is a need for a local agent together with a regional or national firm for wider coverage.
- Subagency. The subagency instruction similarly combines some of the advantages the sole agency and multiple agency structures. A subagent is instructed by the main agent and does not have a direct contractual relationship with the client. The main agent will normally share his commission with the subagent if the purchaser is introduced by them. The relationship between the client and main agent is maintained and the commission rate is normally the same as for sole agency, while increasing the exposure given to the property. The problem can be that without any direct contact between the client and subagent, the subagent can be unmotivated and there can be a lack of control over their conduct.

(c) Fee basis

Fees are normally on an ad valorem (commission) basis which incentivizes the agent to achieve the highest price and means that you will not pay if they are not successful at completing a sale. Where the commission will be a high figure you should consider whether another fee basis would be more appropriate. This could be time based, a fixed fee, or a combination of one of these and an ad valorem fee. Your ability to dictate the fee structure will depend on the type and size of the job and its attractiveness to the agent.

The traditional fee basis for a sole agency arrangement is 10% of the annual rent for a letting and 1–2% of the price for a sale, but the basis should relate to the overall size of the fee and the level of work involved. For multiple and joint sole agency the total fee will normally be higher. The scope to negotiate variations on standard fee arrangements is dictated by the size of the project and the competition between agents. For large instructions substantial discounts can sometimes be achieved.

(d) Managing the agent

Estate agency business is regulated by the **Estate Agents Act 1979**. The Act applies to any activity which has the object of effecting an introduction between potential contracting parties, or of bringing such an introduction to a successful conclusion. This includes both disposals and acquisitions.

Section 18 of the Act provides that on being instructed the agent should confirm your liabilities under the agency contract, including the circumstances in which you will become liable to pay remuneration and the amount of remuneration and any other payments. It is also good agency practice to confirm in writing all the details of the instruction including:

- important information about the property and the interest to be marketed;
- the method of sale and a type of instruction (see below);
- the asking price and terms of disposal;
- viewing arrangements.

The agent should keep you informed of progress on the project and must inform you promptly in writing of any offer received, unless you have specifically indicated that certain offers, for example those below a certain price, need not be notified.

The common law and statute impose various obligations on the agent to act competently and honestly in the interests of the client. They must exercise reasonable skill and care in carrying out their instructions, disclose details of any personal interest or that of any connected person, and follow a set of detailed regulations in relation to any monies held on behalf of clients.

There are also various legal restrictions on the information that can be provided by an agent in the course of marketing and negotiations. The law of contract provides some redress to a purchaser who is induced to enter into a contract on the basis of a **misrepresentation** made by the vendor or the agent, but the purchaser has to demonstrate that they would not have entered into the contract otherwise and that they have suffered a loss. The **Property Misdescriptions Act 1991** imposes a strict liability on estate agents during marketing and negotiations not to include any factual inaccuracy in the information provided to prospective purchasers. Any breach of the Act is subject to criminal proceedings and potentially large fines. Agents are under an obligation to check all information provided, and cannot simply pass on information provided by the client. You can speed up the process by always having documentation ready to support any information you give the agent and by providing access to your solicitor to verify information when necessary.

Throughout the marketing period you should ensure that you maintain control of the process, setting a work plan and timetable at the start of the project, monitoring the agent's activities, and ensuring that momentum is maintained. Commonly the agent will be enthusiastic in the early days of the instruction but may lose some interest over time. You will need to find ways to maintain the level of energy, with regular marketing meetings, and new initiatives, such as a new advertisement, promotion or a change in the terms offered.

As offers are received you will need to consider their merits taking into account the following issues:

- all the terms offered. You should take into account all elements of the offer and use a discounted cashflow model to analyse the value overall (Section 1.2);
- the seriousness of the prospective occupier/owner and the chances of their progressing to completion;
- in the case of assignment or subletting of a lease, the **covenant strength** of the prospective occupier. In addition to ensuring that the landlord will give consent to the assignment or subletting, you need to satisfy yourself that the liability will not revert to you. In the case of subletting you remain liable as the headlessee; in the case of assignment you may continue to be liable after disposal if the privity of contract rule applies to the lease (leases granted before 1 January 1996) or if you have had to guarantee the obligations of the assignee (refer to Section 5.1.8).

You should make full investigations of the proposed occupier's financial position. You should obtain at least three years' recent accounts, references from a bank, a previous landlord and if possible an accountant and one or more suppliers. It is normally advisable to obtain a company search and if possible a credit rating or credit analysis from one of the financial information companies. Apart from needing all this information for your assessment of covenant strength, the landlord is likely to require it in approving the assignment or underletting. It is often advisable to take the advice of a qualified accountant in interpreting this information. If you have serious doubts about financial security you should insist on a financial guarantee, preferably from a bank, or a rent deposit.

5.1.4 Terms of the disposal

Chapter 3 provides a detailed explanation of purchase and leasing terms – refer to this in determining the important issues on disposal. Where you are offering an assignment of an existing lease you will be restricted in the transaction you can offer; unless you are able to negotiate with the landlord and new tenant a simultaneous surrender and re-grant of the lease, the assignee will have to take on the lease as it stands. A **premium or reverse premium** can be used to reflect changes in the market since the original grant of the lease. Additionally, you may have to provide **indemnities** or payments for certain liabilities arising under the lease, such as repairing and reinstatement obligations or service charge liabilities.

Where you are offering a sublease, the interested party may ask you to provide a leasing proposal for a subletting. In this case the process works as a reverse of the situation described in Chapter 2. Any proposal should be prepared with the assistance of the agent and your lawyer. You will need to structure the transaction to meet the needs of the prospective occupier while trying to limit your ongoing liability as far as possible. Refer to Chapter 2 for an explanation of the process of negotiating Heads of Terms and legal documentation and to Chapter 3 for a discussion of the main terms to be considered. These chapters focus on the leasing situation from the perspective of the occupier acquiring the property but much of the advice and information is relevant in the case of a disposal.

In order to limit your ongoing liability you need to negotiate terms and draft the documentation to pass as many as possible of the liabilities on to the new occupier with the minimum residual risk left with you. Where you are negotiating a sale, you should normally be able to discharge all your liabilities under the sale transaction. In the case of subletting this will need to be done through tightly drafting the sublease but you will not normally be able to completely remove the risk arising out of default by the subtenant. In the case of assignment, your original tenant liability will continue under the principle of privity of contract if the lease was granted prior to 1 January 1996, but if the lease was granted after that date the landlord can only require you to guarantee the obligations of your immediate assignee and if they assign your liability will terminate. It is standard for an assignee to provide indemnities which make them liable if the landlord pursues you as original tenant after the assignment. These may not help you if the assignee himself gets into financial difficulty as in that case the indemnities will probably not be of any practical value.

5.1.6 Landlord's consent

You have identified vacant space, found a new occupier to take it over and agreed the terms of the disposal. You may feel that you have already had to jump enough hurdles, but where you hold the premises under a lease you will still need to obtain the consent of the landlord.

Alienation includes assignment, subletting, parting with possession, sharing occupation, charging, and mortgaging. There is no implied covenant against alienation, but modern leases normally expressly prohibit alienation without consent. It is rare to see an absolute bar on alienation, but in this case the landlord will have a free hand to refuse consent. In the majority of cases, the lease provides that assignment, subletting and charging are permitted with landlord's consent. Even if the lease does not specify that consent is not to be unreasonably withheld, (or indeed if the contrary is stated), this is implied by Section 19 of the Landlord and Tenant Act 1927.

When making an **application** to the landlord for consent you should include:

- the terms of the proposed transaction. In the absence of an express provision in the lease, the landlord may not always be able to require to see an agreed underlease or other leasing documents but must be furnished with sufficient particulars to convey the nature of the transaction. Where an agreed or proposed underlease is available it is normally advisable to supply this to the landlord. If this is not possible Heads of Terms should normally be sufficient, but the more detail you can provide the better;
- a full financial information package for the proposed occupier and any sureties. Providing as much information as possible will help prevent delay, and there is normally no advantage in withholding information. The package should include at least three years' recent accounts, any company reports or other financial statements and at least two references preferably from a bank or previous landlord. References should be accompanied by any relevant correspondence, particularly the letters requesting them. Any other information you have available, such as company searches or management accounts, should also normally be provided;
- details of the proposed occupier's trade and the proposed use;
- an undertaking to pay the landlord's reasonable costs in dealing with the consent. This prevents delay as the landlord will not normally proceed until such an undertaking has been given. You should ask the landlord to provide an estimate of the likely costs.

The landlord has an obligation to deal with the application within a reasonable time, under the Landlord and Tenant Act 1988. Time is not considered to run until you have provided the landlord with all the information he can reasonably require, so providing as much information as possible from the start will speed up the process. Additionally, for leases granted after 1 January 1996, any conditions specified in the clause will have to be complied with before time starts to run and under the Landlord and Tenant (Covenants) Act 1995 these do not have to be reasonable (see below). What is considered a reasonable period will depend on the circumstances and the nature of the application. For most commercial transactions you should normally expect to have at least an indication of the landlord's position within 28 days of providing all relevant information, but this will depend on the complexity of the transaction and the other circumstances. The landlord also has a duty to pass on the application expeditiously to anyone else whose consent is required, such as a superior landlord or mortgagee.

If the landlord does not respond within a reasonable timescale, you can adopt one or more of the following strategies:

- obtain a declaration from the court that the alienation should be permitted;
- commence proceedings against the landlord for damages for the delay. You may be able to include any loss of a premium or other financial terms agreed with a proposed assignee or subtenant;
- where you are sure it would be unreasonable for the landlord to refuse consent, you can simply proceed with the transaction;
- in some very limited situations it may be appropriate to seek an injunction forcing the landlord to grant consent.

The last two of these could expose you to a damages claim or forfeiture proceedings if the landlord subsequently shows that consent could have been reasonably refused. In some rare situations the transaction itself could be reversed if you proceed without consent but this would not normally apply if you have applied for consent and the landlord has failed to deal with the application.

If consent is to be granted this will be documented either by a **letter licence** or a **formal licence**. The landlord may seek to impose **conditions**, such as the following:

- conditions relating to financial security, such as rent deposits and guarantees;
- a requirement that the assignee enters into direct covenants with the landlord – this ensures that the new tenant will remain liable if they later assign. This is automatic for leases granted after 1 January 1996, but for these tenancies the assignee is automatically released on assignment unless they are required under the provisions of the Landlord and Tenant (Covenants) Act to guarantee the next assignee (Section 3.4.3);
- a requirement that the sublease is contracted out of the security of tenure and compensation provisions of the Landlord and Tenant Act 1954 (Section 5.2 below);
- a requirement for further subletting to be prohibited in the sublease;
- a requirement for payment of landlord's reasonable costs in dealing with the matter.

Whether such conditions are reasonable will depend on the particular circumstances and legal advice should be taken.

If the landlord refuses consent you will need to identify whether he is entitled to do so. Leases sometimes contain restrictions on alienation, for example prohibiting assignment to unlimited companies or companies which do not meet certain financial criteria, and subletting below market value or below the passing rent. However, these provisions are not always enforceable, depending on the precise wording of the clause and when the lease was granted, and legal advice should be taken. Where such a provision is enforceable you may be able to structure a transaction to accommodate the restriction, for example by paying the tenant a reverse premium to compensate them for a high annual rent.

The extent to which the landlord and tenant can agree in the lease to restrict the landlord's scope to refuse consent to an assignment depend on whether the lease (or agreement for lease) was granted before or after 1 January 1996. Some attempts by the landlord to set out in the lease the grounds on which consent is to be considered will fail, but there is greater scope to do this where the lease was granted after 1 January 1996, as a result of the Landlord and Tenant (Covenants) Act 1995 (Section 3.4.3). As a general rule, for leases granted prior to 1996, consent can only be refused on grounds which the courts define as reasonable, and restrictions stated in the lease are often not enforceable.

As far as the courts are concerned, reasonable **grounds for refusal** depend on the circumstances but some of the following points are of relevance.

- The landlord is entitled to be satisfied that the assignee or subtenant can meet the financial and non-financial obligations of the lease. If, for example, the proposed tenant wishes to use the premises for a use that is prohibited in the lease, or it appears that they will not be able to pay the rent, these may be reasonable grounds for refusal. A common test applied is that the tenant's profits after tax should have been equal to three times the annual rent for three years, but this test cannot be seen in isolation. Any surety, rent deposit or other security being offered must be taken into account. The landlord is in a stronger position to refuse consent on financial grounds where the transaction is an assignment than for a subletting.
- The impact of the transaction on the value of the landlord's interest is of relevance but only if he is likely to suffer a real financial loss (i.e. not just a paper loss), for example if he is anticipating a sale or mortgage. It is only in this situation that the landlord can require that the assignee or subtenant is of equal or better financial standing than the existing tenant because of the impact of covenant strength on the capital value of his interest; in other situations it is enough for you to show that the new tenant will be able to comply with the lease and pay the rent. Where, for example, an assignment to a poorer covenant or an underletting at less than market rent would prejudice the value of the landlord's interest and he is marketing that interest for sale it may be reasonable for him to refuse consent.
- The reason must be directly related to the landlord and tenant relationship concerned. The landlord cannot for example refuse consent to prevent a tenant of other premises owned by him vacating those premises.
- It is usually, but not always, reasonable for the landlord to refuse consent where the tenant has committed a serious and longstanding breach of

covenant. The landlord will normally be expected to proceed with consideration of the application so that consent can be granted once the breach is remedied.

Unless there is an express provision to the contrary, the landlord is not entitled to any fine or premium for the consent, and where the tenant is required to pay the landlord's costs these must not be so large as to represent a fine.

If the new tenant requires a **change of use** or another change which is not permitted by the lease, the landlord will normally have a wide power to refuse consent and will not have to act reasonably in considering the application for consent.

If consent is refused or **conditions** are imposed, the onus of proof is on the landlord to show that the grounds for refusal were reasonable. However, the landlord does not have to show that the grounds were based on fact, only that they were reasonable in the circumstances given the information available at the time. Where the landlord bases his decision on information that subsequently proves inaccurate, this will not make the refusal unreasonable.

Where consent is unreasonably withheld or is granted subject to conditions, the tenant is released from the obligation to obtain consent. In this situation, you can either simply proceed with the transaction or obtain a **court declaration** confirming the refusal was unreasonable. The former course of action is highly risky and could expose you to a damages claim or forfeiture.

There are special provisions relating to **building leases** (other than those granted by government departments or local or public authorities), which are those for more than 40 years involving redevelopment by the tenant. If the lease (or agreement for lease) was granted before 1 January 1996, consent to alienation is not required where it is effected more than seven years before the end of a building lease, regardless of the wording in the lease. Notice in writing of the transaction must be given to the landlord within six months. Where the lease was granted after 1 January 1996, the tenant may have to apply for consent.

5.1.7 Other legal issues

(a) Surrender

A surrender can be effected expressly or by operation of law. It is usually advisable to complete an **express surrender** with a **formal release** from all obligations under the lease. An express agreement to surrender a protected lease at a later date requires approval by the court.

A surrender by **operation of law** entails the tenant offering and the landlord accepting a surrender by conduct, for example by accepting the return of the keys and the lease document.

(b) Offer back provisions

Some leases contain a provision for the tenant to offer to surrender to the landlord before seeking consent to assign. This is sometimes referred to as an **Alnatt provision** after a case in which this type of clause was considered. If the lease is one to which the Landlord and Tenant Act 1954 applies (see below), the offer back provision is ineffective; you will be unable to compel the landlord to consent to the assignment. The provision may be enforceable if the lease is contracted out of the Landlord and Tenant Act or if the Act does not apply for any other reason, but this a legally complex issue and offer back clauses in unprotected leases are sometimes not enforceable. If the clause is not enforceable, there is a possibility that the tenant's alienation rights will be frustrated.

(c) Charging and mortgaging

In some situations you, or a proposed assignee, may wish to use the lease as security for a loan. This is normally where there is value in the lease, usually as a result of a passing rent below the market level – Section 1.4.2 (c). In some cases a lender may wish to place a charge on a lease even if it has no capital value. The charge can give the lender certain rights, such as a right to apply for relief from forfeiture, which may be important to the lender if the occupier's main business is carried on from the premises.

Leases often prohibit charging and mortgaging and you should check whether your lease contains such a restriction. The main reason for restrictions on charging and mortgaging is usually the landlord's concern to avoid assignments by the backdoor; where consent to an assignment is refused the tenant could grant a charge to the new tenant and enable them to take over the lease. Where you can demonstrate that this is not the intention, the landlord may be prepared to approve a charge or mortgage.

(d) Sharing occupation

It is common to find a provision prohibiting sharing of occupation, although sharing with **group companies** is often allowed. The main reason for the prohibition is the possibility of a company sharing on an informal basis obtaining security of tenure. If you wish to allow another occupier to use space in your premises you will need to check the provisions of the alienation clauses and the legal position on security of tenure in the particular circumstances.

5.1.8 Contingent liabilities

On disposal you will need to consider whether you may have any future liability under the terms of the disposal.

This is rarely the case under a freehold disposal, although in some cases you may have to provide an indemnity as part of the transaction for certain costs, such as those associated with a restrictive covenant or with environmental risks.

Where you are disposing of a lease by assignment or subletting there is the possibility that there may be a contingent liability, a potential for liability in the future if an assignee or subtenant defaults. If an assignee is unable or unwilling to satisfy their obligations under the lease, the landlord may be able to look to you as a former tenant. They will not normally be able to pursue you if the lease was granted after 1 January 1996 as the Landlord and Tenant (Covenants) Act 1995 removes the liability of original tenants (Section 3.4.3). However, even in these cases, the landlord will normally be able to pursue you for a breach of covenant which occurred prior to the assignment, for example if he can show the premises were in disrepair before you left them.

If the lease was granted prior to 1 January 1996 (or was granted pursuant to an agreement made before that date), the privity of contract rule applies; if you were the original tenant, you will normally remain liable for the full term of the lease and possibly for any holding over period after expiry. This liability relates to the main financial terms of the lease such as payment of rent and service charge, as well as all the other obligations contained in the lease, such as repairs.

The risk of a future contingent liability is normally greater where the disposal is by subletting as you will retain your obligations under the headlease and your ability to recover from the subtenant will be dependent their financial strength.

You will need to assess the risk of the assignee or subtenant defaulting and where you consider the risk to be material you may need to budget for the cost and to note the contingent liability in your company's financial statements. It is advisable to keep yourself informed of the financial position of assignees and subtenants and to set up a system for obtaining annual reports and company searches at regular intervals to give you early warning of any problems.

If you encounter a claim by a landlord in respect of a lease of which you have disposed, you should take legal advice immediately on how best to resolve the matter. For 'fixed' payments, such as rent and service charge, the landlord must notify you of the claim within six months of the amounts falling due (or by 30 June 1996 for payments due before 1 January 1996) unless proceedings have already commenced, and he will not be able to recover amounts of which he has not notified you within this period. However, he does not need to notify within this six-month period of outstanding payments which are not 'fixed', such as claims for breach of the repairing covenant. Where the former tenant pays in full the amounts claimed he can require the landlord to grant an overriding lease of the premises and may be able to regain possession. The tenant can then use the premises or market them again for a disposal. The same rules concerning notices and overriding leases normally apply for a claim by the landlord against guarantors.

5.2 LEASE EXPIRY

Tenants who occupy property on a lease protected by Part II of the Landlord and Tenant Act 1954 have a **right to renew** the lease on expiry except in certain circumstances. The terms of the new lease are governed by the 1954 Act as are the procedures for renewal.

Landlord and Tenant Act renewals are a complex legal area and there is significant scope to lose your rights under the Act if the procedures are not properly followed. It is important that you take legal advice in good time and all notices should normally be served by a solicitor. The following sections outline the main elements of the regime to enable you to take an informed approach, but in order to ensure you protect yourself properly professional advice should be taken.

5.2.1 Protected tenancies

Where the following circumstances apply the tenancy will be protected by the Part II of Landlord and Tenant Act 1954.

- Tenancy. There must normally be a tenancy for a fixed term exceeding six months. Licences are not protected. A number of types of tenancy are specifically excluded, such as agricultural holdings, mining leases, and service tenancies.

  Where a tenancy is granted for a fixed term of less than six months it will be protected if there is provision for renewal. Therefore, periodic tenancies of, for example, a week, a month or a year, are afforded protection.

  A lease will not be protected by the Act if it was expressly contracted out by agreement of the parties approved by the court when it was granted (Section 3.4.2).
- Business occupation. The premises must be occupied by the tenant for the purpose of a business. Occupation involves both control and physical occupation and it is a matter of fact and degree whether in a given case a tenant is considered to occupy.

  Business is defined as 'any activity carried on by a body of persons whether corporate or unincorporate' and includes any trade, profession or employment. The definition covers, amongst other uses, offices, shops, garages, warehouses, factories, laboratories, hotels, cinemas and surgeries.

  Where the premises are used partly for business and partly for another use, such as where a professional uses part of his home as an office, the Act will prevail, as long as the user clause in the lease allows for business use.

Where a tenancy is one to which the Act has applied at any time during its contractual term, the 1954 Act will apply. Therefore where the tenant is not in occupation at expiry but has been previously the procedures discussed below must be followed.

5.2.2 Procedures

If a lease is protected by the Landlord and Tenant Act 1954, it will not automatically expire at the contractual expiry date. After the contractual term has ended, the tenancy continues until it is determined by one of the following methods.

- Termination by the landlord. The landlord can serve a **Section 25 notice** any time between six and 12 months before the contractual expiry date. The notice must specify the date on which the tenancy is to terminate, which can be any date from six to 12 months away but no earlier than the contractual expiry date. The notice can be served to take effect after the contractual expiry date, in which case the lease will continue until the date specified in the notice for expiry. The landlord must specify in the notice whether he is prepared to grant a new tenancy.

  If the landlord is not prepared to grant a new tenancy, he must state one or more of the listed grounds for refusal:
  - breach of repairing obligations;
  - persistent delay in paying rent;
  - other substantial breaches;
  - the landlord will provide suitable alternative accommodation for the tenant;
  - possession is required for letting or disposing of the property as a whole (where the tenancy part of a larger holding);
  - the landlord intends to demolish or reconstruct;
  - the landlord intends to occupy the premises (where the landlord has held his interest for more than five years).

  The landlord cannot rely in subsequent court proceedings on any ground not stated in the Section 25 notice. It is therefore common for landlords to include a number of grounds some of which they may choose not to rely on later.

  Once the landlord has served a Section 25 notice, the tenant must serve a counter notice under Section 25(5) within two months to preserve his right to apply for a new tenancy. If the tenant fails to do so he will lose the right altogether. Once the tenant has served the counter notice the procedures described below for application to the court for a new tenancy come into play. Proposed reforms of the lease renewal procedures include withdrawing this requirement for a tenant's counter notice.
- Termination by the tenant. If you wish to vacate the premises or you wish to renegotiate the terms on which you occupy, you will wish to terminate your current tenancy.

  If you do not wish to renew a protected tenancy you must serve a **notice to quit**, sometimes known as a **Section 27 notice**. If this is served three months or more before the contractual expiry date, it will take effect on that date. If it is served after the date three months prior to the contractual expiry date, the tenant must give at least three months notice expiring on a **quarter day**. The quarter days are 25 March, 24 June, 29 September and 25 December. The tenant loses all rights under the Act once a valid Section 27 notice takes effect.

From the date 12 months before the expiry of the contractual term you have the right to serve on the landlord a **Section 26 request** terminating the contractual tenancy and requesting a new one. Again, the date specified for the request to take effect must be between six and 12 months ahead and not earlier than the contractual expiry date.

A tenant who holds under a periodic tenancy (such as a monthly tenancy) can serve a notice to quit but cannot serve a Section 26 request for a new tenancy and if he wants to obtain a new lease under the Act he must wait for the landlord to initiate the Landlord and Tenant Act procedures with a Section 25 notice.

If the landlord wishes to oppose a new tenancy he must respond within two months of the tenant's Section 26 request with a counter notice specifying the grounds of opposition. Again the landlord will be restricted to these grounds in the subsequent proceedings.

The Section 25 notice (or Section 26 request) must be served by (or on) the person who is defined as the '**competent landlord**'. No problem arises when the tenant's landlord is the freehold owner but where there is a chain of tenancies the landlord to whom the rent is paid may not be the competent landlord for the service and receipt of notices under the Act. In such a case the appropriate landlord is the person who has a reversion (legal interest after expiry of the sublease) of at least 14 months duration. If the immediate landlord holds under a tenancy with longer than 14 months to run or a shorter tenancy which is protected by the Act where no notice has been served terminating that tenancy within 14 months, then the immediate landlord will be the competent landlord. If, on the other hand, the immediate landlord does not hold a lease with more than 14 months to run or protected by the Act, or has a protected tenancy in respect of which a Section 25 notice or Section 26 request has been served, then the superior landlord or the next in the chain holding an interest of longer than 14 months will be the competent landlord.

Once a notice under Section 25 or Section 26 has been served the landlord can apply to court for an **interim rent** to cover the holding over period until the grant of a renewal tenancy. At the moment the tenant cannot apply for an interim rent (although there are proposals to change this) so sometimes in a falling market the landlord will be content to let the tenant hold over for some time and it may be up to you to initiate termination procedures. In this case, it is normally advisable to make your application to the court as soon as possible in order to expedite the court order for the new tenancy at the market rent. The interim rent will be payable from the date on which the application was made to the court, or the date for termination of the tenancy specified in the landlord's Section 25 notice or the tenant's Section 26 request, whichever is the later. The interim rent is a market rent assuming a year-to-year tenancy disregarding the items listed below at 5.3.3(c). The interim rent is usually approximately 10–20% below the market level for a fixed-term lease, but this depends on market preferences at the time.

Section 40 of the Landlord and Tenant Act 1954 provides a mechanism for the landlord and tenant to establish ownership and occupation. The landlord can serve a notice requesting details of occupation and subtenancies; the tenant can serve on the landlord a notice requesting details of the landlord's interest, length of tenancy and so on. These notices may be served at any time from two years prior to the expiry date or other date at which the tenancy could be terminated. The information must be provided within one month of receipt of the notice but there is no special penalty for non-compliance.

If the landlord and tenant both wish to negotiate a new tenancy negotiations can initially commence out of court. The tenant cannot apply to the court for a new tenancy until two months after service of the Section 25 notice or Section 26 request, but the application must be made within four months. If the landlord opposes the new tenancy or the parties cannot reach agreement on terms the application will be decided by the court.

5.2.3 Terms of the renewal tenancy

The court will not deviate from the terms of the current tenancy unless the party requesting a change can provide justification for it. In practice, most lease renewals are agreed between the parties and the court does not have to intervene. However, the terms a court will be likely to grant form the framework for the negotiation as the parties can revert to the court if necessary. The main terms which the court may change in the renewal lease are as follows.

- Property demised. If the parties do not agree on the area to be demised, the court will determine this. The tenant cannot insist on inclusion of parts of the current demise that are not occupied by him for business purposes (such as those sublet), but the landlord can require the court to order that the demise should include the whole of the original letting.
- Length of term. The parties are free to agree upon whatever length of term they choose but the court is limited to a maximum term of 14 years. The court will take into account all the circumstances, including:
  - the length of the current tenancy;
  - the nature of the business;
  - the age and state of the property and the prospects of redevelopment;
  - any intentions the landlord may have to redevelop.
- Rent. In the absence of agreement, this will be determined by the court as the rent the holding would reasonably be expected to let at in the open market by a willing lessor, having regard to the other terms of the tenancy.

  Certain matters must be disregarded when assessing the new rent, including goodwill resulting from the tenant's business, the effect on rent of your occupation and the effect of improvements to the premises which you have carried out.

  The rent will normally be negotiated between the landlord's and tenant's surveyors based on comparable evidence adjusted to reflect the attributes of the property and the terms of the lease. In the event of disagreement the court will decide based on the evidence adduced by the parties. Sometimes the rent cannot be finally agreed or determined until all the other terms of the tenancy are decided as these may affect the rent.

The rent review pattern may follow that of the current tenancy, but the frequency will be based on market practice at the time of the renewal. Where the existing lease does not contain reviews the court may include them in the new tenancy and these may be freely floating (i.e. upwards or downwards).

Once the terms of the new tenancy have been determined by the court and an order for the new tenancy is made, the tenant has 14 days in which to refuse to accept the terms. The lease will commence four months from the court's determination unless the parties agree an earlier date for commencement.

5.2.4 Compensation for disturbance and improvements

(a) Compensation for disturbance

Where the landlord is successful at opposing a new tenancy under the Act as a result of one of the non-fault grounds (such as the landlord's wish to redevelop or occupy), the tenant may be entitled to compensation for disturbance.

Compensation is calculated by reference to a formula relating to the rateable value (RV) of the premises. There are two bases, an old and a new one, relating respectively to the old rateable value (as at 31 March 1990) and the new Uniform Business Rate rateable value in the list at the date of the landlord's notice, as shown in Table 5.2. The old basis is used where the landlord's notice opposing renewal (Section 25 notice or counternotice under Section 26) was served on or before 31 March 1990, and the new one where it was served after that date.

For business premises with a residential element where the lease was granted prior to 1 April 1990 in respect of which the landlord's notice opposing renewal was served between 1 April 1990 and 31 March 2000, tenants can opt to apply the old rateable value with a multiplier of 8 or 16. To do this the tenant must serve a notice during the period for the application to court (i.e. 2–4 months

Table 5.2 Compensation multipliers

	Old basis (landlord's notice on or before 31 March 1990)	*New basis (landlord's notice on or after 1 April 1990)*
Tenant in occupation less than 14 years	Old RV × 3	New RV × 1
Tenant in occupation more than 14 years	Old RV × 6	New RV × 2

after the Section 25 notice or Section 26 request), so it is important to consider the matter in good time to decide which basis is most advantageous.

In some cases you may have to proceed with an application for a new tenancy even if agreement has been reached with the landlord to vacate, in order to protect your right to compensation. This might occur if the landlord includes a number of reasons for refusal in the Section 25 notice, including some relating to the fault of the tenant (such as non-payment of rent). In the absence of an express agreement with the landlord obliging him to pay you compensation, you will only be certain of retaining your right to compensation if the court refuses to grant a new lease on the basis of one of the non-fault grounds.

If you come to an agreement with the landlord out of court to vacate where the landlord wishes to redevelop or occupy himself, you should ensure that the compensation you would expect to receive if the court upheld the landlord's refusal is reflected in the terms agreed.

An agreement in the lease or other document excluding the tenant's right to compensation may be void.

(b) Compensation for improvements

Where the tenant has undertaken alterations which enhance the value of the landlord's holding, and notified the landlord prior to making the alteration, they may be entitled to compensation. This will be the lower of the cost of the works and the increase in the value of the landlord's interest. The claim must be made within three months of the end of the lease. Refer to Section 4.6.3 for further details.

5.2.5 Unprotected tenancies

If you hold premises under a lease which is not protected by Part II of the Landlord and Tenant Act 1954, you will not have any statutory right to renew on expiry. Your right to occupy ends on the expiry date and there is no need for you to serve a notice to quit, unless there is a notice provision in the lease. The landlord can obtain a court order for possession and force you to leave the premises.

If the landlord allows you to stay on after expiry, the terms of your occupation will depend on the circumstances. You should always take legal advice to confirm the basis and ensure you protect yourself effectively. In most cases where you remain in occupation you will hold under one of the following types of tenancy:

- a tenancy at will – in this case the landlord is entitled to possession immediately he gives you notice to vacate and there is no security of tenure;
- a licence to occupy – this applies if you are occupying pending agreement on terms for a new lease and again you will not have a right to renew;

- a periodic tenancy – the period of the tenancy will depend on the period by which rent is calculated in the lease. If you continue to pay rent and the landlord accepts rent on the same terms as the lease, the term of the periodic tenancy will be the period by which the rent is calculated. If, as is usual, the rent is expressed as an annual sum, the tenancy will be an annual periodic tenancy and the landlord and tenant will have to give six months notice to determine it. If the landlord allows you to occupy on this basis you may acquire security of tenure; it is therefore important that you obtain legal advice to ensure that any rights of this kind are protected.

Whether occupation after expiry of a contracted-out tenancy is on the basis of an implied tenancy at will or periodic tenancy will depend on the circumstances. In particular, the length of time you occupy and whether the landlord accepts rent or any other payment will have an effect on the way the arrangement is interpreted.

If the landlord takes proceedings to recover possession a tenancy will not normally be created and you will simply be considered to be **holding over** under the expired lease. In this case rent will not be payable but the landlord instead will be entitled to **mesne profits**, which is a payment of damages for use and occupation of the premises. This will normally be the market rental value but the rent reserved under the lease will often be taken into account when calculating this. In some cases the landlord may be entitled to double rent, such as when a tenant fails to vacate after having given notice to quit.

It is important that you take legal advice in good time to ensure you can protect yourself properly. The consequences of not doing so could include having to relocate at short notice, losing any renewal rights you may have acquired or an action for double value if you hold over after expiry. You may also need to instruct a surveyor or commercial agent. In the early stages the surveyor will advise on comparative occupancy cost levels for the existing property and alternatives to ensure that decisions are based on accurate information. In the later stages he can advise you and negotiate with the landlord.

5.2.6 Lease expiry – timetable and tactics

In view of the importance of premises to the operation of the business, lease expiry should be planned for in good time. If you are going to have to relocate on or soon after expiry, you should be starting the acquisition process described in Chapter 2 in earnest at least one year before the relocation date. If there is a chance that you will need to acquire a building which requires significant construction works, for example where the supply of the type of buildings you will require is limited and you may need to look to a pre-let or development situation, you may need to start the first stages of the acquisition process two or more years before expiry.

Where tenants wish to remain in occupation they sometimes wait for the landlord to serve a Section 25 notice and fail to take advantage of their right to bring the tenancy to an end and initiate proceedings for a new tenancy. Where the rental value of the premises is lower than the passing rent is it usually advisable to terminate the tenancy as soon as possible. Where the market has risen it may be best to sit tight, but the uncertainty of holding over will affect your ability to plan.

Table 5.3 shows an indicative timetable for action in connection with a lease expiry. The actions and dates shown are best practice in many cases, but will depend on the particular circumstances.

5.2.7 Break options

(a) Tenant's break option

There will normally be notice provisions in the lease which will have to be followed in order to exercise the option. If your lease is protected by the Landlord and Tenant Act, the contractual notice will normally be enough to satisfy the Landlord and Tenant Act, but if the contractual notice period is less than the three months required by Section 27 of the Act, you may need to serve an additional notice.

If the lease is protected by the Act, it has been argued that the break clause could be used to negotiate a new lease with the possibility of using your security of tenure rights to get a lease at market rent. However, recent case law suggests that the courts will not uphold the request for a new lease so you take the risk that you will be left with no right to occupy. You will need to consider this risk and take legal advice when deciding whether to adopt this strategy.

Table 5.3 Lease expiry timetable

Date	Action
Two years before expiry	Consider relocation on expiry – assess business needs and evaluate options (Chapter 2). Instruct lawyer to advise on legal position and surveyor to advise on the terms of a new lease. Consider meeting with the landlord to discuss his plans and if appropriate commence discussions on the terms of a new lease. Consider whether a section 26 request should be served at the earliest date, i.e. 12 months before expiry.
One year before expiry	Decide whether you wish to stay or relocate. Speak to landlord for an update on his plans. If you wish to relocate or there is a chance the landlord may oppose a new lease, commence the acquisition process for alternative premises (Chapter 2). Serve Section 26 request (if appropriate).
Six months prior to expiry	Last date for service of Section 26 request to take effect on expiry.

If you hold a lease which is contracted out of the security of tenure and compensation provisions of the Landlord and Tenant Act, you will only have to serve any contractual notice specified in the lease to exercise the break. You can operate the break simply as a means of renegotiating terms but you will not have a right to renew and may have to vacate if terms cannot be agreed. It may be difficult to negotiate a market transaction as the landlord will know that the costs of relocating will motivate you to stay and this will enhance his negotiating strength.

If a tenant's break option is **conditional** on performance of any covenants, you will need to address all relevant obligations in good time, to ensure the condition can be met. A conditional covenant carries the serious danger that you may lose your right to break (even over trivial breaches), and you need to do all you can to ensure that this does not happen. The main problem is that you will not always know whether you are in breach of covenant, so you need to clarify the position as far as possible with the landlord. You cannot force the landlord to tell you whether he considers you are in breach. However, if you do all you can to comply with the provision the courts will often be inclined to uphold your right to break even if you are technically in breach.

The first step is to write to the landlord approximately nine months before the break date asking him to notify you formally of any breach of covenant and any action he considers you should undertake before the break date to ensure you are not in breach. The main items are likely to be dilapidations and reinstatement (Section 5.3 below). Once you have the landlord's response you should comply with all his requirements. You will not have the option of leaving work undone and making a payment of damages if there is a conditional break clause as you would have normally. If you dispute any of the landlord's requirements, ensure that you resolve the matter in good time to carry out any necessary work before the break date. If you are aware of any other items that would constitute a breach, ensure these are remedied, even if the landlord does not refer to them. You should take into account any obligations that will arise towards the end of the term, such as an obligation to redecorate in the last three months.

If the landlord does not respond to your request to notify you of any breaches, it is advisable to instruct a building surveyor to prepare a schedule of dilapidations. You should notify the landlord that you consider the works listed to be sufficient to discharge your liability, with a request that he confirm. You should also review the lease, possibly with the assistance of your lawyer, to determine whether there are any other breaches that should be remedied. This strategy should minimize any risk of forfeiting the right to break, but it may not eliminate it altogether and legal advice should be taken so that you can decide on whether to make commitments on another property.

(b) Landlord's break option

If the landlord has an option to break and the lease is inside the Act the landlord will have to serve a Section 25 notice and will have to prove one of the approved grounds for refusal of a new tenancy if he wishes to obtain possession. If the notice period specified in the lease is longer than six months, the landlord has an extended period for serving the Section 25 notice. For example, if the contractual notice period is 15 months, the landlord can serve the Section 25 notice up to 21 months before the date on which it will take effect, but it must take effect on or after the date the contractual notice does. Once the landlord serves the Section 25 notice, you will be able to apply for a new lease in the usual way and can hold over in the meantime under the terms of the old tenancy, subject to any interim rent.

If the lease is contracted out of the Act, the landlord will have full scope to terminate the tenancy on the break date and can obtain an order for possession if you refuse to leave.

5.3 DILAPIDATIONS AND REINSTATEMENT

The obligation to **repair** applies during the currency of the lease, but in practice is usually addressed when the tenant exits at expiry or earlier determination of the lease. Refer to Section 4.5.1 for a discussion of the legal issues relating to repairing covenants.

The normal practice for commercial leases is for the landlord's building surveyor to inspect the premises a short time before or after expiry and prepare a **terminal schedule** of dilapidations. In some cases, if the landlord has reason to believe the premises are being left to deteriorate substantially, he may serve an **interim schedule** of dilapidations during the term, combined with a Section 146 notice as a precursor to forfeiture proceedings (Section 4.13.1). In either case, the schedule details all items of disrepair and normally includes an estimate of the cost of undertaking each item. The total cost includes fees, VAT and loss of rent for the period of the works after expiry of the lease and this forms the basis of the landlord's damages claim.

The loss of rent element compensates the landlord for rent he would have received during the period the repairs are being undertaken. In some cases the figure is significant, and it is often advisable to address the dilapidations issue early in order to bring forward the repairs, ideally to before expiry, and reduce or eliminate the loss of rent.

The normal practice following service of a terminal dilapidations schedule is for the landlord's and tenant's surveyors to negotiate, both in respect of the items included and the cost attributed to each of them. In most cases the claim will be agreed between the parties, but if necessary the landlord can use one or more of the remedies available to him for breach of covenant generally (Section 4.13.1). In practice the matter is usually addressed after expiry and therefore remedies which only apply during the currency of the lease, such as forfeiture, are not available. The normal approach is to pursue an action in damages.

The landlord's claim will be limited by the restrictions on the tenant's obligations provided by Section 147 of the Law of Property Act 1925 (internal decorative repair) and the Leasehold Property Repairs Act 1938 (other repairs) – see Section 4.5.1. Additionally, under the Landlord and Tenant Act 1927, the landlord's claim will be limited by the amount by which the disrepair diminishes the value of his reversion. In particular, where the landlord undertakes major construction works or **redevelopment**, he will not normally be able to support a damages claim in respect of disrepair.

You will normally need to instruct a surveyor to act for you in dealing with a dilapidations claim. Normally both a valuation and building surveyor will be required but sometimes one or the other will suffice. In some cases legal advice is also necessary. The RICS has a recommended fee scale for dilapidations work based on the amount claimed, but a time-based fee is more common.

In addition to dilapidations you may have an obligation to reinstate any **alterations** carried out when you moved in or during your occupation. Your obligation in this respect will depend on the wording of the alterations covenants in the lease and any licence to alter. If there is no express **reinstate-**

If the landlord is redeveloping you may be able to escape liability for dilapidations.

ment obligation the landlord cannot normally require you to reinstate. Where there is such an obligation there may be an exclusion in respect of certain improvements or there may be an obligation on the landlord to notify you prior to expiry which alterations he requires you to reinstate. Even if there are no such limitations on your obligation, you should liaise with the landlord in good time to determine which items he requires you to reinstate, as he may be prepared to let some remain if they benefit the premises.

Once you vacate the landlord will only be able to enforce the reinstatement covenant with an action in damages. He will not normally succeed in a claim for reinstatement of items which enhance the capital value of the premises, which are those which would benefit the average tenant in the market for the premises. These will usually be confined to enhancements to services and base building fixtures, and in some cases facilities such as kitchens and toilets. Your professional advisers may be able to limit your liability by negotiating to exclude items from the dilapidations and reinstatement obligation, but this will depend on the terms of the repairing and alterations clauses in the lease and any licences to alter.

The cost of dilapidations and reinstatement will need to be estimated in advance for internal management accounting and for the financial analysis used for decision making. When the move-versus-stay option is considered, you need to factor in these costs. A quantity surveyor or building surveyor will normally be able to give you a broad indication of the costs without significant work.

The tenant normally has a right to remove **tenant's fixtures**, but this will depend on the wording of the lease and any other document, such as a surrender deed. Tenant's fixtures are defined by reference to, amongst other things, the extent and purpose of annexation to the building and whether they are physically capable of removal without causing substantial damage, so you may not be entitled to remove all items which you have installed. The right to remove tenant's fixtures will normally expire when you give up possession, but it can expire earlier. You will normally have an obligation to make good any damage caused by removal.

5.4 COMPULSORY PURCHASE

Public sector bodies (including privatised industries) have a wide range of rights to acquire property compulsorily if it is needed for certain purposes, including infrastructure or certain types of development scheme. This may affect you whether you are an owner occupier or you occupy under a lease. If you encounter a situation in which an authority wishes to acquire your property compulsorily, it is important that you manage the process to protect your rights and secure the maximum amount of compensation.

5.4.1 The process

A confirmed Compulsory Purchase Order will give the acquiring authority the right to enter, use and acquire land. The notice period for entry can be as little as two weeks. You should seek to use the statutory process to maintain control and avoid serious disruption to your business. This can be done by negotiating with the acquiring authority throughout the process. The aim is to understand what the authority's objectives are, and to explain your concerns. Often the needs of both parties can be met.

You can identify an intention to compulsorily acquire land in a number of ways. These include press announcements, a letter from the acquiring authority or a request to provide details of ownership.

Once you discover a scheme you should exert pressure immediately. It is easier to influence the form, timing or even existence of a scheme in its early days. It can be more difficult as the scheme matures. When any **Compulsory Purchase Order** is made (submitted to the Secretary of State for confirmation) there is an opportunity to object within a defined timescale. It is usually advisable to do so – it will put you in a better negotiating position. Compulsory purchase orders and the schemes underlying them are not set in concrete but if you are not one of the objectors, you cannot expect to influence the result. By objecting you may be able to have the scheme abandoned or altered, to delay or speed up the acquisition, or to agree compensation in advance.

If there are objections, the Compulsory Purchase Order has to be confirmed after a **public inquiry**. Deals can often be struck which could not be done at any other time in the process. At the public inquiry the merits of the scheme and objections to it are heard by an inspector who reports to the Secretary of State for the Environment. The outcome of the inquiry may be confirmation of the Compulsory Purchase Order, confirmation with amendments or rejection. Once the Order is confirmed, after a short period for legal challenge, the acquiring authority will have powers to acquire compulsorily, subject to formalities.

5.4.2 Assessing compensation

If the Compulsory Purchase Order is confirmed and your property is acquired, the rules for assessing compensation will normally be those contained in the Compulsory Purchase and Compensation Acts (particularly the Land Compensation Act 1961 and 1973 and the Compulsory Purchase Act 1965), although these are sometimes adjusted.

There are six main rules normally used for assessing compensation (known as **The Six Rules**).

1. No allowance shall be made on account of the acquisition being **compulsory**.
2. Compensation is for **open market value** assuming a willing seller.
3. Compensation will not take into account any value attributable to a **special**

suitability of the land for a particular use where that use could only be applied pursuant to statutory powers or where there is no market except for the acquiring authority.

4. Any value relating to **illegal or unlawful uses** (including planning contraventions) will not be compensated.

5. Where there is no general demand or market for the property, compensation may be assessed on the basis of the reasonable cost of reinstating an **equivalent building**.

6. Compensation may include **disturbance** payments. These may include removal costs, loss of profits, fitting out costs and the costs of finding and surveying new premises.

In addition to these Six Rules the following principles apply.

- The scheme itself has to be ignored. For example, any decline in the surrounding area due to the scheme, or the shape of the site if distorted by the scheme, have to be disregarded.
- If the property has development potential there are certain **planning permissions** which can be assumed.
- Compensation may include a payment for diminution in the value of land caused by **severance** (cutting off part of the site) or by **injurious affection** (the construction and use of the scheme).
- Even if only part of your property is required for the scheme you may be able to force acquisition of the entire property.

If your property will be required by a scheme but the authority has not yet commenced the acquisition, its value may be affected in the meantime. You may be able to force the authority to acquire it by using a **blight notice.** This requires them to purchase your interest at normal compensation values. The mechanism can only be used in respect of properties with a rateable value less than £18 000. There are further provisions for compensation in some cases where a property is not to be acquired but is severely affected by the scheme.

Refer also to Section 4.9.5 where compensation for planning decisions is discussed.

5.4.3 Advisers

Compulsory purchase advice can be obtained from specialist surveyors and lawyers. If you are to protect yourself properly, it is important that you are advised throughout the process by a professional with an understanding of both how to assess compensation and how to employ the procedures effectively. Good negotiation skills are also important as the adviser will normally act on your behalf in negotiations with the authority. In many cases a surveyor will be able to deal with the matter, but in more complex cases a lawyer will also be required.

There is a scale of charges for surveyors fees, known as **Ryde's Scale**, which is the basis that acquiring authorities tend to use in assessing compensation for fee expenses. Most professionals undertake this type of work on a time basis and you may have to suffer a shortfall in fees recovered. Much of the work is consultative in nature and therefore a time based fee is normally appropriate, but other fee bases can sometime be negotiated.

FURTHER READING

Denyer-Green, B. (1994) *Compulsory Purchase and Compensation*, The Estates Gazette, London.

Freedman P. and Shapiro, E. S. (1994) *Commercial Lease Renewals: A Practical Guide,* Macmillan.

Male, J. M. (1995) *Landlord and Tenant*, Pitman Publishing, London.

Murdoch, J. (1993) *The Estate Agents and Property Misdescriptions Acts,* The Estates Gazette, London.

Smith, P. F. (1993) *The Law of Landlord and Tenant*, Butterworths.

Stephens, S. (1981) *The Practice of Estate Agency*, The Estates Gazette, London.

Appendix A
Measurement for valuations

This appendix is based on the RICS code of measuring practice, the copyright of which is owned by the Royal Institution of Chartered Surveyors.

In many valuations, whether it be a formal capital or rental valuation, an informal calculation of the rent that should be paid under a lease, a rent review valuation, or an insurance valuation, the valuation is based on a calculation of the floor area of the building. To ensure a consistent approach to measurement the RICS and ISVA jointly publish the *Code of Measuring Practice*, currently in its fourth edition. This is accepted as the basis for agreeing floor areas. It provides core definitions of area as follows.

Net Internal Area (NIA) – the usable area measured to the internal face of the perimeter walls at each floor level. It excludes toilets, lift rooms, tank rooms, stairwells, liftwells, permanent lift lobbies, and areas used in common or for the purpose of essential access. Areas under 1.5 m (5 ft) in height are separately stated along with car parking areas.

This is the most widely used definition and is the common basis of measurement for estate agency, valuation, rating and property management of offices, some business premises (combined office and light industrial use) and retail premises.

Gross External Area (GEA) – the area measured externally at each floor level. It excludes open areas such as open balconies, fire escapes and parking areas but all internal areas are included.

This definition is used for town planning and Council Tax banding and some rating measurement.

Gross Internal Area (GIA) – the area measured to the internal face of the perimeter walls at each floor level. It excludes perimeter wall thicknesses and external projections, external open-sided balconies, covered ways and fire escapes but includes columns, piers, stairwells and internal common areas.

This is the common basis for measuring for building cost estimation, and for estate agency, valuation, rating and property management of industrial buildings and some light industrial premises.

Appendix B
Lease summary checklist

Premises:
Lease date:
Lease summary prepared by: Date:

Item	Details	Lease clause
Current landlord		
Current tenant		
Original/former tenants or landlords		
Guarantors/other parties		
Term and expiry date		
Break options – landlord/tenant		
Expiry/break notice		
Security of tenure?		
Areas and items demised		
Other rights of access and use e.g. car spaces, lifts, staircases etc.		
Rent:		
Original rent reserved		
Current rent payable		
Rent payment dates		
Paid in arrears or advance?		
Rent review:		
Rent review pattern		
Next review		
Review notice procedure		
Time of the essence?		
Arbitration/independent expert?		
Notice procedure		
Basis of review, main assumptions and disregards		

Service charge:
Reserved as rent?
Service charge year
Current budget
Apportionment basis
Special exclusions
Service charge payment dates
Landlord's services
Hours of access/service

Insurance:
Reserved as rent?
Cover placed by landlord – buildings? other?
Cover placed by tenant – contents? fixtures and fittings? alterations?
Tenant to reimburse?
Rent cesser/break option
Insurance payment dates

VAT:
Landlord elected? Tenant liable to pay?

Interest:
Chargeable on rent? service charge? insurance? other?
Interest rate payable
Reserved as rent?
Days grace?
Charge from due date?

Other charges:
Electricity? Gas? Water?
Landlord's costs for breach?
Other?

Rates:
Tenant or landlord to pay?
Rateable value
Appealed?
Current rates payable

Repair:
Standard of repair
Decorations
Exclusions?
Dilapidations – notice procedures

Alterations:
Permitted with consent?
Reinstatement obligation
Consents granted

Use:
Permitted use
Change allowed with consent?
Permitted hours of use

Alienation:
Assignment/subletting permitted with consent?
Restrictions?

Remedies for breach:
Does privity of contract apply?
Forfeiture? Grace period?
Tenant's right of set off?

Special provisions/obligations:
Any special terms?

Other documents   Date/parties
Agreement for lease
Licence to alter/letter licence
Deed of variation
Supplemental lease
Other documents

Appendix C
Addresses and information sources

Note: This appendix aims to provide a starting point for readers looking for information or services relating to commercial property. Inclusion in the list does **not** represent a recommendation and readers should make their own investigations of any services provided. The list is not exhaustive and readers should not assume that the organizations listed are the only source of the information or service.

OCCUPIERS' ORGANIZATIONS

British Retail Consortium
Bedford House, 69–79 Fulham High Street, London SW6 3JW, tel. 0171 371 5185, fax 0171 371 0529
IDRC Europe
Postbus 59366, 1040 KJ Amsterdam, The Netherlands, tel. 0031 20 301 2237, fax 0031 20 301 2202
NACORE Europe (The North American Association of Corporate Real Estate Executives)
Saxon Court, 502 Avebury Boulevard, Milton Keynes MK9 3HT, tel. 01908 692812, fax 01908 692813
The Property Managers Association (PMA)
c/o Pizza Hut, 1 Imperial Place, Elstree Way, Borehamwood, Hertfordshire WD6 1JN, tel. 0181 732 9000, fax 0181 953 1909

PROFESSIONAL AND TRADE ORGANIZATIONS

Association of Planning Supervisors
15 Rutland Square, Edinburgh, EH1 2BE, tel. 0131 221 9959, fax 0131 221 0061

British Association of Hotel Accountants (BAHA)

193 Trinity Road London SW17, tel. 0181 672 6444

The British Council for Offices

The College of Estate Management, Whiteknights, Reading RG6 6AW, tel. 01734 885505, fax 01734 885495

The British Council of Shopping Centres

The College of Estate Management, Whiteknights, Reading, RG6 6AW, tel. 01734 885505, fax 01734 885495

British Institute of Facilities Management

67 High Street, Saffron Walden CB10 1AA, tel. 01799 508608, fax 01799 513237

The British Property Federation

35 Catherine Place, London SW1E 6DY, tel. 0171 828 0111, fax 0171 834 3442

Chartered Institute of Management Accountants (CIMA)

63 Portland Place, London W1M 4AB, tel. 0171 637 2311, fax 0171 631 5309

Incorporated Society of Valuers and Auctioneers (ISVA)

3 Cadogan Gate, London SW1X 0AS, tel. 0171 235 2282, fax 0171 831 4390

Institute of Chartered Accountants (ICA)

Chartered Accountants Hall, Moorgate Place, London EC2P 2BJ, tel. 0171 920 8100, fax 0171 920 0547

Institute of Management Consultants

Fifth floor, 32–33 Hatton Garden, London EC1N 8DL, tel. 0171 242 2140, fax 0171 831 4597

Institute of Revenues Rating and Valuation (IRRV)

41 Doughty Street, London WC1N 2LF, tel. 0171 831 3505, fax 0171 831 2048

Law Society

113 Chancery Lane, London WC2A 1PL, tel. 0171 242 1222

National Association of Estate Agents

Arbon House, 21 Jury Street, Warwick CV34 4EH, tel. 01926 496800

Royal Institute of British Architects (RIBA)

66 Portland Place, London, W1N 4AD, tel. 0171 580 5533

The Royal Institution of Chartered Surveyors (RICS)

Surveyor Court, Westwood Way, Coventry CV4 8JE, tel. 0171 222 7000, fax 0171 334 3800

Royal Town Planning Institute (RTPI)

26 Portland Place, London W1N 4BE, tel. 0171 636 9107

Society of Business Economists

11 Bay Tree Walk, Watford WD1 3RX, tel. 01923 237287

DIRECTORIES OF CONTACTS

Directory of Industrial and Commercial Property Contacts, Newman Books, London.

Estates Gazette Directory is published in the first issue of each month of *Estates Gazette* (see below) and provides details of surveyors, estate agents and other property organizations throughout the country.

The Surveyors' 500, Tuckers' Directories and National Law Tutors, Manchester.

The Top 1000 Law Firms and All Barristers' Chambers, Chambers & Partners, London.

Waterlows Solicitors' and Barristers' Directory, Waterlow Information Services, London.

Prichard, J. *The Legal 500*, Legalease, London.

PROPERTY MARKET INFORMATION AND ANALYSIS

Applied Property Research (APR)
97 St John Street, London EC1M 4AS, tel. 0171 251 5654, fax 0171 250 1079
Barber White Property Economics
Fitzroy House, 18–20 Grafton Street, London W1X 4DD, tel. 0171 408 1366, fax 0171 499 6279
Focus
Ingram House, 13–15 John Adam Street, London WC2N 6LD, tel. 0171 839 7684, fax 0171 839 1060
Investment Property Databank (IPD)
7–8 Greenland Place, London NW1 0AP, tel. 0171 482 5149, fax 0171 267 0208
Property Market Analysis (PMA)
Tower House, 8–14 Southampton Street, Covent Garden, London WC2E 7HA, tel. 0171 379 5130
Segal Quince Wicksteed Limited
The Grange, Market Street, Swavesey, Cambridge CB4 5DG, tel. 01954 231931
Roger Tym & Partners
9–10 Sheffield Street, London WC2A 2EY, tel. 0171 831 2711, fax 0171 831 7653

A number of the surveying firms publish property market performance statistics and undertake research and consultancy work.

PROPERTY PERIODICALS

Estates Gazette, *Estates Times* and *The Property Week* all run local market surveys and legal and general interest articles.

Other property related journals include:

Architects Journal

Architectural Review

Corporate Real Estate Executive (published by NACORE)

Europroperty

Premises and Facilities Management

Property Law Bulletin

Property Review

The Valuer

London Office Guide is published three times per year providing details of available space in Greater London.

BENCHMARKING

BWA Premises Management

Kings House, 32–40 Widmore Road, Bromley, Kent BR1 1RY, tel. 0181 460 1111, fax 0181 464 1167

The Centre for Interfirm Comparison

Capital House, 48 Andover Road, Winchester SO23 7BH, tel. 01962 844144, fax 01962 843180

Hillier Parker

77 Grosvenor Street, London W1A 2BT, tel. 0171 629 7666

The Operational Property Databank

7 Greenland Place, London NW1 0AP, tel. 0171 482 5149, fax 0171 267 0208

Pall Mall Services Group

125 Acre Lane, London SW2 5UA, tel. 0171 274 8622, fax 0171 737 7448

PIMS

Seventh floor, Moor House, 119 London Wall, London EC2Y 5ET, tel. 0171 628 1155, fax 0171 628 2455

Price Waterhouse

1 London Bridge, London SE1 9QL, tel. 939 3000

Procord

2 The Briars, Waterberry Drive, Waterlooville PO7 7YH, tel. 01705 230500, fax 01705 230501

Savills Commercial

20 Grosvenor Hill, Berkeley Square, London W1X 0HQ, tel. 0171 499 8644, fax 0171 409 8748

Index

Printed in Great Britain
by Amazon

57894250R00169